Springer Japan KK

H. Deguchi

Economics as an Agent-Based Complex System

Toward Agent-Based Social Systems Sciences

With 91 Figures

Springer

Hiroshi Deguchi
Department of Computational Intelligence and Systems Science
Interdisciplinary Graduate School of Science and Engineering
Tokyo Institute of Technology
4259 Nagatsuta-cho, Midori-ku
Yokohama 226-8502, Japan

ISBN 978-4-431-67965-3 ISBN 978-4-431-53957-5 (eBook)
DOI 10.1007/978-4-431-53957-5

Printed on acid-free paper

Originally published by Springer-Verlag Tokyo in 2004
Softcover reprint of the hardcover 1st edition 2004

springeronline.com

SPIN: 10986967

Preface

This book is a condensed and revised version of the original Japanese edition that was published in 2000 by Union of Japanese Scientist and Engineers (JUSE), Tokyo. The purpose of this English version is to introduce a new research program in the area of agent-based modeling (ABM) and its application to socioeconomic systems.

In the last decade, several ABM research programs have been developed around the world. In Japan, since the early 1990s some groups have been exploring an agent-based approach in the social sciences. In the area of agent-based modeling the main research programs have developed from distributed artificial intelligence (AI) or complex adaptive systems (CAS). One of the most important influences on agent-based modeling comes from the CAS paradigm, with a focus on learning and the evolution of agents.

At the same time, other research programs in the area of complex systems with decision makers were being developed by two different scientific approaches. One comes from the study of dissipative structure and synergetics as an extension of statistical physics. From the very earliest stages, ABM has been influenced by the concept of self-organization, structural change as bifurcation of dynamical systems. However, genetic algorithm and the evolutional approach became more popular, partly because they do not require the construction of a dynamical systems model to explore an agent-based approach.

The other research program developed from traditional systems sciences that includes cybernetics, operations research, general systems theory, and the soft systems approach. Systems scientist H.A. Simon and his colleagues have developed a computational approach at Carnegie Mellon University and other places. In particular, K. Carley and her group have developed a mathematical and computational organization theory and an agent-based approach at Carnegie Mellon. At the University of Michigan, R. Axelrod and M. Cohen started their research program under the influence of J.H. Holland, who is a founder of the CAS paradigm. In Europe, N. Gilbert and the Simulating Society (SIMSOC) group also devised a simulated approach in the social sciences in the early 1990s.

In Japan, Y. Takahara and his colleagues, who with M.D. Mesarovic were founders of a mathematical general systems theory, have developed a hierarchical systems and a mathematical organization theory. I am also a pupil of Y. Takahara and can state that in Japan the research program of ABM has been influenced by this mathematical general systems approach.

Hard systems approaches such as cybernetics or operations research provide problem-solving tools and methods under the given aims of systems. In the social sciences we have to start from a definition of the problem in human activity systems. In other words, systems sciences provide epistemic devices to comprehend human activity systems as ideal types.

P.B. Checkland and his group have developed a soft systems methodology. This approach gives a framework to contain and solve the problems of human activity systems. It became popular in the United Kingdom and Japan but not in the United States. Gaming simulation is a different soft research program for complex systems with decision makers. ABM also must be combined with an interpretative approach such as gaming simulation or soft systems methodology for comprehensive real-world grounding. In the early stages of our research program of agent-based modeling in Japan, gaming simulation was introduced as an essential tool for real-world grounding. In this book we focus on gaming simulation.

In ABM the stress is on a bottom-up approach that is strongly connected to a kind of reductionism. If we are interested in designing socioeconomic systems, this approach must be combined with a macro functional system.

In this book we present a research program for a socioeconomic system as a science of autonomous agents with deviations from the standard research programs of neoclassical economics. Traditional economists are not interested in ABM as a research program, because they adhere to strong academic conventions for treating rational agents. They cite the early works of the repeated prisoner's dilemma by R. Axelrod again and again but never cite his recent work with ABM, as it deviates from their paradigm. In the last decade they have tried to expand their framework of rational decision making, and evolutional game theory is a typical development. More recently they have introduced learning theories such as reinforcement learning into their framework. Nevertheless, the gaps between these different paradigms are deep and wide. It would seem that to effectively analyze a competitive multiagent system and its institutional design we require a new paradigm. In this book we construct a research program for a socioeconomic system as a science of autonomous agents, with deviations from the standard research programs of neoclassical economics or standard rational decision theory. In ABM the focus is very much on agent-based simulation, as simulation is an important tool for ABM. We also use agent-based simulation in this book with an emphasis on the mathematical foundation of ABM. We introduce two original mathematical frameworks, a theory of dynamical social learning and a descriptive theory of economic exchange among agents. We focus on the theoretical, mathematical, and methodological aspect of ABM.

In Chapter 1 we set forth the basic methodology for ABM as used throughout this book. Repeatability and refutability are important criteria for scientific theories. In soft systems methodology the focus is also on the traceability of the analysis. A computational approach satisfies these three methodological requirements in the context of simulation, but computational formulation does not give mathematical formulations and theorems even when grounded in the Keep It Simple and Stupid (KISS) principle. A computational model might become valid under the docking method of modeling, but it does not directly provide interpretive or other ways of real-world grounding. We investigate the epistemological foundation of ABM.

In Chapter 2 we investigate the dynamic decision-making model of social science and its bifurcation analysis that provides a mathematical foundation for some ABM. The bifurcation model provides hard-core theory for structural change and self-organization. In this chapter we also focus on the concept of indirect control where we control the boundary conditions of a system to manage the steady state of the system by introducing social learning dynamics (SLD). SLD provides a framework for structural change that creates the theoretical framework of ABM simulation. We also introduce the concept of indirect control of socioeconomic systems by using the SLD framework to manage structural change and its stability.

In Chapter 3 we apply SLD to the problem of norm formation and the collapsing process. The model was originally formulated and analyzed by R. Axelrod in the form of agent-based simulation as a norm game and meta-norm game (Axelrod, 1997). We include a reconstruction of a norm and meta-norm game by SLD and also provide other models of SLD in socioeconomic systems.

In Chapter 4 we focus on the fact-finding ability of agent-based simulation, introduce agent-based simulation with reinforcement learning, and show the emergence of interesting strategies in the problem of the tragedy of the commons. We also compare agent-based simulation with gaming simulation. We introduce the concept of indirect control into the gaming simulation framework and show how to improve the tragedy of the commons.

In Chapters 5 and 6 we have provided a mathematical foundation for a national economy from a descriptive point of view. In Chapter 5 we give the mathematical basis for a bottom-up construction of an economic state. We include an axiomatic foundation for a state description of economic exchange, in which we introduce the concept of a minus state of stock quantity with an input–output description. As traditional structuralism theory states, we have abstracted a deep axiomatic structure, known as exchange algebra, from the repeated surface structure of a bookkeeping system. The System of National Accounts (SNA) and the System of Social Demographic Statistics (SSDS) are both important in giving a descriptive foundation of economic and social systems.

In Chapter 6 we give a mathematical reconstruction of the SNA from a bottom-up point of view. Macroeconomic variables for the stock and flow of a national economy are formulated from our exchange algebra.

Chapter 7 provides an application of gaming simulation on a small-scale model of a national economy known as a virtual economy. This virtual economy is constructed using the framework explained in Chapters 5 and 6. We also try to combine agent-based simulation with gaming simulation.

In Chapters 8 and 9 we give an application of social learning dynamics for analyzing the lock-in process of network- or platform-oriented industrial structures. We extend the concept of network externality to the concept of platform externality, and we treat the dynamic change of the utility of agents under the platform structure of industry.

In conclusion, the chapter construction flows as follows:

```
                        → Chapter 3 → Chapter 8 → Chapter 9
                        |
Chapter 1 → Chapter 2 → Chapter 4 → Chapter 7
                        |                 ↑
                        → Chapter 5 → Chapter 6
```

Agent-based social systems sciences have started to consolidate their long research history. In the United States the North American Association for Computational Social and Organizational Science (NAACSOS) was established in June 2003 (www.dis.anl.gov/naacsos/home.html). In Japan the Pacific Asian Association for Agent-based Social Sciences (PAAA) was also established in June 2003 (www.paaa.econ.kyoto-u.ac.jp), and in September 2003 the European Social Simulation Association (ESSA) was established (http://essa.cfpm.org). These associations are connected with one another and will contribute much to the design of our global society in the twenty-first century.

I hope this book will make a valuable contribution to the field and will provide a conceptual, theoretical, and practical framework to aid the study, understanding, and planning of new solutions that meet the challenges of our rapidly changing world.

I express my gratitude for the support provided by the Japan Society for the Promotion of Science in the form of 2002/2003 Grants-in-Aid for Publication of Scientific Research Results.

I am also very grateful to the staff of Springer-Verlag Tokyo, and to Mrs. Masayo Whitaker and Mr. Andi Whitaker for their helpful editorial support.

This book is dedicated to my beloved wife Fusae and to the memory of my parents, Shigeru and Kimiko Deguchi.

Hiroshi Deguchi, DSc, DEc
December 2003
Tokyo, Japan

Contents

Chapter 4
The Tragedy of the Commons and Its Agent Based and Gaming Modeling

Chapter 5
State Space Complexity and Exchange Algebra

Chapter 6
The National Economic Field and Algebraic Reconstruction of SNA

Chapter 7
Gaming Simulation and the Dynamics of a Virtual Economy

Chapter 8
Network Oriented Industrial Structures

Chapter 9
Learning Dynamics in Platform Externality

Chapter 1
Systems Description and Models in Economics

Purpose of This Chapter

In this chapter, we discuss the outline of the basic design of this book and we make it clear why this new economic theory needs to be investigated.

We also discuss why the study of complexity in social sciences must become a science of autonomous agents. This chapter describes a research program of economics as a science of autonomous agents.

The topics of this chapter are developed further in the following chapters.

In this chapter, a methodological issue for modeling a complex system with decision makers is also summarized.

The science for autonomous agents needs to develop a research program different from the one used in natural science to investigate universal laws.

An economic system is regarded as a system of autonomous agents in this chapter.

The methodological issue for this purpose is examined from a variety of perspectives such as system recognition, agent based simulation and the control and management theory of agents.

1.1 A Systems Description of Economics Systems

In this book we shall investigate the new frontier of socioeconomic systems. Adam Smith, David Ricardo and others gave the foundation of classical economics in the 18th century. Economics has developed under the marginal revolution in the 19th century, and the Keynesian revolution in the 20th century along with the growth of a capitalistic economy.

Economics, as the core of social science, takes pride in having contributed to the capitalist economy and in creating elegant mathematical tools for modeling economies. Nevertheless when the model is examined from the methodological viewpoint we feel a strange sense of incongruity with the theory. This sense of incongruity arises from the fact that social science intends to clarify systems, which include different types of persons and organizations as decision makers.

The first question we have to ask here is what economic variables actually indicate. In macroeconomics several variables that characterize a national economy are used such as national income, investment, consumption, and savings. Richard Stone, a disciple of John Maynard Keynes, systematized the variables into a system of national accounting.

From the theoretical point of view these macro variables are composed of various micro variables that represent the activities of individual economic agents. If these microstates of such economic agents are given as accounting information then the relation between organizational accounting and national accounting must be investigated [Mattessich, 1995a,b].

In other words we need to formulate a micro macro link to describe the theoretical state of an economic system. In the case of hard science a formulation for a state-space is required as a prerequisite for describing a theoretical law in physics. In physics, both micro and macro state-space are required for a systems description. Micro state space is used for an atomic scale description of systems or field description. Macro state space is used for our daily life scale description. Besides, this state-space has an internal structure consisting of a physical quantity.

For example, in the kinetic theory of gases, n $\times$ 6 variables are used for the description of the position and momentum of n molecules of gas. Temperature, pressure and volume are used as a variable for the macro scale description of gas. Statistical mechanics for the kinetic theory of gases has developed as a micro state-space, and thermodynamics has developed as a macro state-space. Statistical mechanics is able to give a foundation for thermodynamics, but we also need to include macro variables of gas such as temperature and pressure from micro state space.

The micro-macro relation between variables in economics can be compared to the case of statistical mechanics. Moreover we have to pay attention to the relation between micro decision mechanisms and macro laws in economics. Contemporary economic theory tries to establish the micro foundation of macroeconomics by using the optimization theory of decision making as a competitive equilibrium for the market in microeconomics. This analysis of the relation between the laws of micro and macroeconomics is essential in the social sciences. Thus in the social science we have to focus on two types of micro-macro relation, one is the relation between state variables and the other the relation between laws of decision and dynamics.

In economics the former means how the state variables of macro-economics can be decomposed into transactions among individual economic agents.

In economics they do not focus on this kind of reductionism approach with the exception of input-output analysis and social accounting matrix [Pyatt, 1985]. Economists usually focus on the market mechanism.

Because they believe that from the theoretical point of view production and distribution of goods are adjusted by market price and Parato optimal state is achieved from first theorem of welfare economics.

1.2 Paradigm Shift

After the Industrial Revolution we had to seek a market control mechanism within a now newly differentiated and complex world. As a result the "invisible hand of God" was proposed as an autonomous distributed control mechanism. The realization that market price can function as a coordination variable for decision-making was a splendid discovery.

The market price was indispensable in an economy where information processing ability was low. Even now we follow the adjusting mechanisms developed at the time of the Industrial Revolution. However, now we have powerful information processing abilities and various types of management mechanisms can be introduced into inter organizational structures.

In the old economy J.M. Keynes introduced macroscopic control mechanisms and the emergence of macroeconomics means the emergence of macro management tasks for government and a consequent need for more information about economic systems.

The System of National Accounts (SNA) was constructed for this purpose. SNA and input-output analysis give anatomical data of the real economy, which has increasingly become important for government policy. Nevertheless there is little connection between the microstructure of accounting transactions of firms and the macro statistics of the SNA. The purpose of this book is to give a firm mathematical foundation for this connection.

On the other hand the role of organization is ignored in classical microeconomics, except from the viewpoints of transaction cost and the acquisition cost of information. Contract theory or agency theory has been developed from an economist's point of view [Salanie, 1998].

Nevertheless the understanding of organization is simple and trivial from the management science point of view. With the coming of the Internet revolution our business environment is becoming global and more complex. More and more complicated interactions and business models are emerging among business organizations. We need to analyze the creation and evolution of these various types of business modes.

We need to understand the diversity of the real economic games played within actual economic systems. Thus the central doctrine of the new economics should be to put an emphasis on the analysis of the variety of the games executed by various types of autonomous agents and the institutional systems design of such economic games.

Modern economics has acknowledged the Pareto optimum principle as an important welfare criterion and now we need to add a variety of new criterions. How to maximize the innovation rate? How to avoid the lock-in of a market? How to treat common properties as free software on the Internet? How to treat intellectual property rights? These are serious questions in our contemporary global and information oriented economy and we have to answer these questions. A new social welfare criterion for economic systems should be created taking into account the changes brought about by the information and communication revolution. An economic theory for a systems design need to be developed.

For this purpose we need to develop new types of decision theory beyond the rational and optimization theory of decision-making. Of course we have a long history of the theory of bounded rationality, but the economic interpretation of bounded rationality is different from the interpretation offered by the organizational scientists such as H.A. Simon [Simon, 1982].

This stress on Nash refinement as bounded rationality leads to a degenerated interpretation.

At the present time there is an underlying paradigm shift from rationality to learning in economics, this learning shift is not a strong movement at this moment, but we maintain that it is a very important and growing trend in contemporary economic theory. In this book we will introduce the social learning dynamics of this learning shift and will construct a new research program of economics and decision theory based on the advances of the new information and communication revolution in our modern society [Kuhn, 1970; Lakatos, 1983].

We will clearly demonstrate a research program of economics as a complex system, consisting of autonomous agents. It is not a complex system such as dissipative structure or chaos found in the natural sciences. It is not the same as the paradigm of a complex adaptive system such as artificial life or a genetic algorithm, but our research has led along these lines.

In this book we will demonstrate a new research program of economic systems from an agent based, process and communication oriented point of view. We need to analyze the agent society, consisting of autonomous agent interacting with each other. We naturally assume that the agents have rules within which they act. The agent also has a model of their environment and other agents, known as an internal model, and makes use of information to guide his activities. We also assume an ability to learn rules and create their internal model [Wonham, 1976; Rosen, 1974A,B, 1985]. The agent society that we are interested in consists of several types of functional differentiated agents. In other words we assume a micro-macro linkage in the agent society.

1.3 Conception of This Book

We shall introduce a new management method for agent society by controlling the institutional boundary conditions of the society.

In this chapter we discuss the methodology of agent-based approach in social sciences. We describe a mathematical framework in Chapter 2. In Chapter 3 we introduce social learning dynamics and discuss mutual and hierarchical commitment and also centralized and hierarchical commitment in an agent society. We analyze norm formation and collapsing processes with examples. In Chapter 4 we introduce agent-based simulation with a classifier system. We analyze the problem of the tragedy of the commons, which is a typical social dilemma, in agent based and gaming simulation. We compare both approaches.

In Chapter 5 we pay attention to the micro state space of the economic agent. We abstract an axiomatic formulation of economic exchange from a bookkeeping system. This algebra gives the micro state space of an economic agent. We reconstruct SNA, system of national accounting, by using this algebra in Chapter 6. This is a construction of macro economic variables from the micro economic variables of agents. In Chapter 7 we analyze a small national economy as an agent society, consisting of nine agents, agriculture, a flour milling industry, a bakery, a steel and machine industry, a government, a bank and central bank and households. This is called a virtual economy. We analyze this virtual economy by using gaming simulation and genetic algorithm. In the last chapter we apply our framework developed in this book to the problem of contemporary economic systems.

In this book, we describe an economic system as an agent society, in which autonomous agents refer and learn mutually. We will construct a research program of economic systems analysis, by which we analyze the diversity and variety of agent activity under various institutional boundary conditions. We analyze these issues and explore contrasts found in the mathematical model, agent based simulation and with human agents, also the contrast between theory and methodology, the contrast between the micro and bottom up approach and the macro and functional approach or the contrast between state description and rule description.

As a result we propose a new methodological framework for agent based modeling and various types of models for analyzing agent society.

1.4 Methodological Issue

In this chapter, the methodological issue of modeling a complex system with decision makers is summarized. We introduce the concept of a poly agent system so as to understand the system of autonomous agents. We consider the system that includes many autonomous agents, but for statistical analysis we consider a system with less agents [Kijima, 1996].

The term complex system has been used for more than a hundred years. In the history of complex systems science we have studied the systemic properties, which are difficult to treat with modern hard science such as Newtonian dynamics [Kampis, 1991]. In this book we focus on social and economic complex systems and their agent based modeling.

Working systems scientists know that the word “complex system” does not denote a fixed research domain but refers to a research frontier of complexity [Takahara, 2003; Rosen, 1978]. In other words the term “complex system” has not been used as a proper name to mean a fixed systemic property, but used as a common name to include several systemic properties in social systems theory, cybernetics and system science [Archer, 1995, 1996; Buckley, 1967]. We focus on some systemic properties of complexity and depending on its research stage or paradigm, our systemic properties of complexity change along with the research program of complexity.

At a certain historical research stage of systems sciences composition and decomposition are important systemic properties of complexity, at another stage they focus on the hierarchy of a system [Mesarovic, 1970]. Cybernetics developed as a theory to analyze the complexity of a purposive system [Ackoff, 1981; Ashby, 1956]. Self-organization, emergence, learning and evolution are also important systemic properties of complexity.

We pay attention to the four major breakthroughs in the theoretical study of complex systems. Then we discuss our research program.

(1) The first one is the cybernetics revolution. In cybernetics we formulated the concept of feedback. The concept of feedback makes it possible to use the word "purpose" as a scientific concept from an engineering point of view.

(2) The second breakthrough comes from computer science. The concept of program plays an essential role in the study of complex systems. In complex systems we focus on hierarchical emergent properties at the higher levels of the systems. The concept of program provides the basis for a non-physical description of the systems law of the higher levels of complex systems.

H.A. Simon stressed the science of the artificial and characterized the concept of "role" as programmed procedures. The object-oriented description of complex soft systems comes from this paradigm [Simon, 1957, 1979].

(3) The third revolution occurred in the area of statistical physics. The paradigms are called synergetics or self-organization that characterize the change of structure by mathematical bifurcation theory of non-linear dynamical systems [Haken, 1998; Nicolis, 1979; Yam, 1997; Zeeman, 1977].

(4) New paradigm, the fourth breakthrough will emerge in the area of agent based modeling and complex adaptive systems [Axelrod, 1997; Carley, 1994]. Multi agent systems are systems in which several agents interact.

The "agent" works to accomplish their goals through the activities of commnication, cooperation, competition, negotiation, transaction, control and management.

Now we focus on social and economic complex systems with multi autonomous agents as decision makers, where we try to clarify systemic properties such as the bottom up construction of systems, indirect regulation of agent societies, learning and adaptation of agents, emergence of function and structure in agent societies, self-organization and structural change of agent societies, etc.

This approach is called ABM (Agent Based Modeling) or the agent based approach. Agent based modeling partly comes from the CAS (complex adaptive system) paradigm. The ABM research program on social science started simultaneously at different places in the 1990s.

In the 1990s several research groups started their research programs in the area of social sciences, such as SIMSOC (Simulating Society) group in Europe [Gilbert, 1994, 1995], ABCE (Agent Based Computational Economics) group at Iowa State University, Center on Social and Economic Dynamics at the Brookings Institution [Epstein, 1996], Computational Analysis for Social and Organizational Systems (CASOS) group at CMU and CAR project at University of Michigan. In Japan the Poly-Agent research group started its activity early in the 1990s [Takagi, Kijima, Deguchi (eds.), 1995].

The reason why we used the term "poly agent" is that we wanted to treat hetero and functionally differentiated agents in addition to the activity of homogenous agents. A concept of the poly agent system is not the research program, which has a hard core of clear mathematics, but a rather looser research initiative. Nevertheless, it is significant for social science to develop a research program in this field. We pay particular attention to the following topics in my research program on ABM.

(1) We focus on social learning rather than rational decision-making. Thus we introduce social learning dynamics and its bifurcation analysis. By using the social learning dynamics we analyze mutual and hierarchical commitment processes among agents.
(2) We will construct a national economy from the bottom up point of view. For describing an agent based economic state of stock and flow we give an algebraic abstraction of a bookkeeping system, which is called exchange algebra.
(3) We also focus on real world grounding by gaming simulation (GS).
In ABM we use computer simulations in which many software agents are included. GS gives validity and real world grounding for ABM simulation.

1.5 Three Types of Realities

We focus on social complex systems that contain autonomous agents and their indirect control principle.

In a social complex system we cannot assume any universal law that rules the system.

For analyzing such a system we distinguish three layers of reality within the system such as theoretical, computational and socially constructed realities.

(1) The first is a theoretical reality where the mathematical models are developed.
(2) The second is a computational reality based on the facts on the simulation. Our digital life world is constructed by computer-mediated communication based on the computational reality.
(3) The third layer of reality is a socially constructed reality. In this layer we pay attention to the subjective and intersubjective construction of reality.

For designing our extended life world a form of reality analysis is used in the area of management science. As complementary approaches to the hard systems approach, Soft systems methodology and constructivism are examples of these kinds of soft and subjective approaches [Bennet, 1980; Espejo, 1996; Flood, 1991; Checkland, 1985, 1990]. Gaming simulation is another approach toward a construction of reality [Arai, 1998; Greenblat, 1981, 1988; Klabbers, 1996]. Now we pay attention to gaming simulation.

Our methodology for a social complex system is shown in the following Fig. 1.1.

We investigate several types of agent societies depending on different levels of realities. We also discuss why the agent-based approach is essential for supporting a community. We also use learning and evolutionary methods for describ-

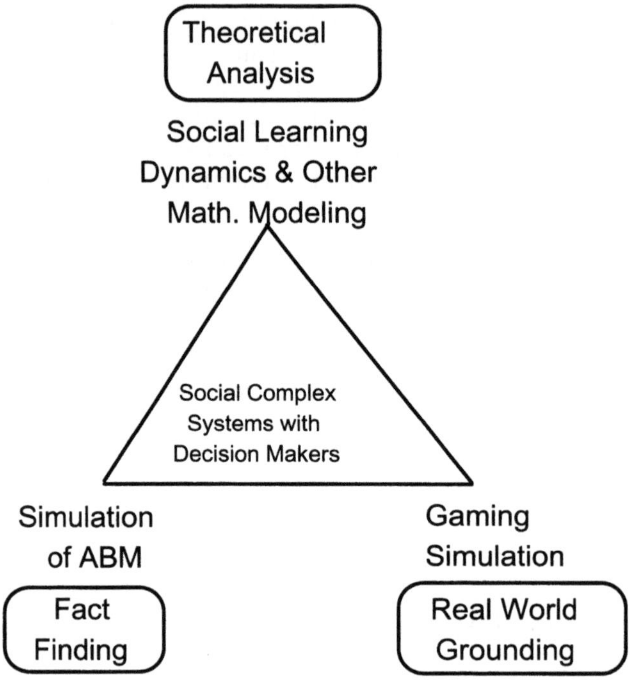

FIG. 1.1 Methodological Triangle

ing software agents. Software agents act as real decision makers. We analyze social, economic and organizational systems by using ABM.

ABM is now developing as an important tool for policy and decision making and for designing tools for human activity systems. Artificial society, the economy and markets are also areas of application for ABM.

On the other hand GS has a long tradition in the fields of education, policy making and design tools. In GS we make a model of various situations of social and economic activities in which many human agents are included as players.

GS and ABM have almost no connection between the two research communities except in Japan. In Japan the two communities interact with each other and have developed an original methodology. We believe that GS is useful for ABM because GS gives validity and real world grounding for ABM simulation. ABM is also useful for GS because ABM makes it possible to construct more complex and agent aided gaming simulation. Gaming simulation with human player (Player Character: PC) and software agents (Non Player Character: NPC) create new frontiers for the gaming of actual society and economic systems.

Uniform agents with non-strategic optimization behavior and the representative agent theory characterize the old paradigm of decision making. The new paradigm of decision making is characterized by agents who learn and mutually refer to internal models and learning of strategy. The central doctrines of the new research program are shown as follows.

(1) Analysis of Various Types of Games of Autonomous Agents.
(2) Design of Institutional Boundary Condition for Indirect Control of Agent Society.
(3) Fact Finding by Agent Based Simulation.
(4) From ABS to Mathematical Model for Mutual Commitment Model of Social Learning.

1.6 Beyond the Systems View of Natural Science

From the viewpoint of natural scientific systems it is assumed there is a universal law. They believe that the variety of each system comes from the variety of boundary conditions of the universal law.

In the theory of dissipative structure and synergetics, description start from the premise of a universal law. This type of systems view is shown as follows (Fig. 1.2).

For an example of the knowledge structure of natural science we will use the case of Ohm's law and Maxwell's law in electromagnetic theory. Ohm's law ($V = IR$) is an example of a macroscopic phenomenological law, which is reduced to Maxwell's law. Ohm's law is a law of specific phenomena. In contrast, Maxwell's law is a universal law governing electromagnetism and enables us to explain a variety of phenomena. Maxwell's equations include the formula $i = \kappa E$ which is also called Ohm's law in the Maxwell equations. It is not identical to the classical Ohm's law such as $V = IR$.

The identification of these two laws becomes possible after constructing a certain relation between the different theories.

Microscopic Universal Law		
↓		
Specific Law Characterized		
Boundary Condition	→Construction→	Macroscopic
of Universal Law	←Reduction→	Phenomenological Law
↓		↓
Microscopic	→Construction→	Macroscopic
State Space	←Reduction→	State Space

FIG. 1.2 A Systems View of Natural Science

Maxwell's law

↓

$i = \kappa E$

Steady State Condition of Maxwell's law	→Construction→ ←Reduction→	Ohm's law (V=IR) Phenomenological Law
↓		↓
i , E: Microscopic State Space	→Construction→ ←Reduction→	I, V, R: Macroscopic State Space

FIG. 1.3 Relation Between Ohm's Law and Maxwell's Law

We assume that the classical Ohm's law is defined by such special conductors with the length as L, the force of electric current is parallel with the conductor and the area of cross-section of the conductor is S. As the electric current density is constant this means a steady state condition for Maxwell's law.

Under these boundary conditions we construct the following macroscopic variables I, V, R from the microscopic variables of Maxwell's law.

(1) $I = \int_S i \cdot n dS$

(2) $V = -\int E \cdot dL$

(3) $R = L/(S\kappa)$, $i = \kappa E$

I, V and R are macro variables, which are used in the phenomenological Ohm's law.

From the steady state condition the following properties hold.

$$I = S|i| = S|\kappa E|, V = |E|L$$

Thus $V = |E|L = S|\kappa E|L/(S\kappa) = IR$ holds.

This means that the classical Ohm's law can be reduced to Maxwell's law. The relation is shown as follows (Fig. 1.3).

Classic control theory is based on such a natural scientific systems view.

However, it is impossible to explain multi agent systems with decision makers by using this type of natural scientific view. For this type of multi agent system we cannot assume a universal law for the system.

1.7 An Epistemological Framework for a Poly Agent System

A social complex system, a poly agent system, consists of a number of autonomous agents. Autonomous agents as decision makers have a variety of activity rules, which change by learning, evolution and the emergence of rules.

It is usual to find heterarchy by functional differentiation in agent society and we will construct management and control principles for this type of agent society.

The society has a certain macroscopic functional requirement. We introduce a new control concept of agent society called indirect control or indirect regulation, where we control the boundary conditions of an agent society to achieve the macroscopic purpose of the society.

In traditional economics, little attention is given to analyze the dynamics of rule based agents and the indirect control of agent society. On the contrary they assume rational and uniform agents who maximize their utilities under perfect or complete information. To protect the theory they do discuss bounded rationality or incomplete information, but actually they are not interested in agent society, which consists of autonomous agents who act according to various strategic rules and from their learning.

Of course the assertion that the Pareto optimum state is obtained by market competition is essential for managing a modern society. They insist that the rational allocation of resources is guaranteed by market equilibrium and the activity of profit maximization.

In the information oriented society we have to focus on the more complex activities of agents. There are many new phenomena on the network and global markets these days. For example platform externality and lock in of the market is becoming very important and it is discussed in Chapter 9.

We need to analyze the activities of a rule-based agent and their systemic properties such as dynamical learning, evolution and the emergence of rules, mutual commitments among agents and the institutional design of boundary conditions for the activities of agents.

For this purpose we introduce the concept of indirect control. For characterizing the indirect control concept we introduce social learning dynamics and its bifurcation analysis on agent society in Chapters 3 and 4. Social learning dynamics are a population dynamics for the multi agent system, which represents the selection of alternatives and its learning. For describing agent based systems it is natural to assume that there is no universal law. Instead we assume a variety of decision making rules and sets of alternatives. An agent selects an action plan from alternatives depending on decision rules, the micro state of the agent and the macro state of agent society. As a result of the execution of the selected plan, the micro and macro state of the society will change. We also need to consider the learning mechanism of decision rules.

An agent selects a desirable plan from alternatives and they might also commit to other agents to change the selection of other agents. A central authority or manager, who plays as a macro functional agent, tries to find various structural parameters as institutional boundary conditions to achieve a better social state. This regulation of structural parameters is called indirect control of an agent society. Mutual commitment among agents is also a form of indirect control mechanism within agent society.

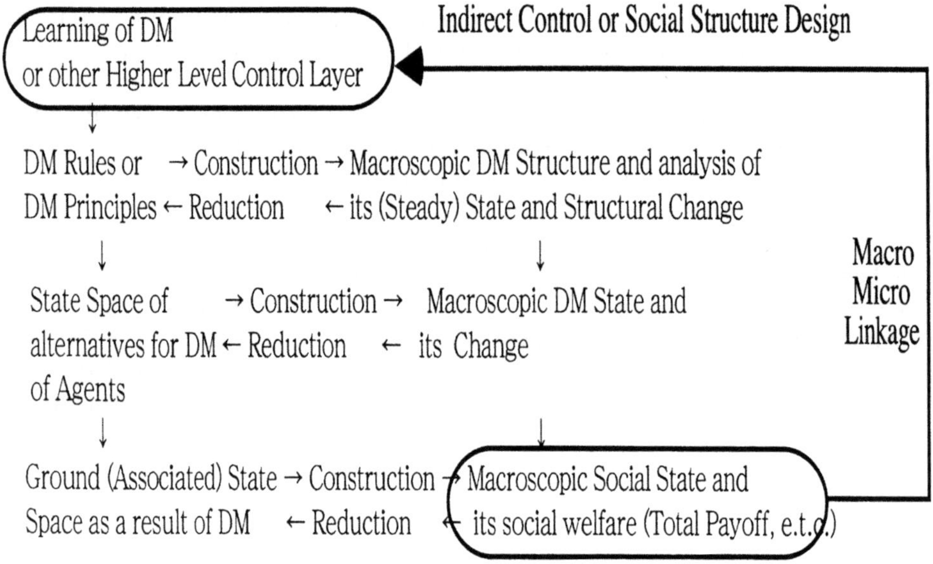

FIG. 1.4 General Modeling Framework for Agent Societies

We can show our methodological framework for systems recognition and a description of an agent based system in the following Fig. 1.4 [Deguchi, 1998, 2000a,b]. We explain our modeling framework for a concrete mathematical model by using social learning dynamics and its bifurcation theory in Chapters 2 and 3.

Chapter 2
Mathematical Research of Evolution and Structural Change
—Social Learning Dynamics and Its Bifurcation—

Aim of This Chapter

In this chapter we focus on dynamical systems and their structural change. We summarize existing models of structural change such as self-organization, evolution, adaptation, and learning.

Within our context we also make a dynamical systems model of autonomous agents. First we show that structural change is analyzed by the bifurcation of a dynamical system. Then we deduce social learning dynamics from social interaction, which can express social learning processes among agents.

We also summarize genetic algorithms and a theory of reinforcement learning. We discuss the concept of internal models, which is a key systemic property of agent learning through these approaches.

2.1 Structural Change and Bifurcation in Social Sciences

We summarize a dynamical systems model for self-organization or structural change and bifurcation analysis. In modern dynamical systems theory and its application we consider that the stable equilibrium state of the dynamical system represents the state of the system. We have to distinguish structural stability from state stability such as asymptotic stability. For analyzing structural change of the system we focus on structural stability. If the pattern of equilibrium points and the trajectories of the system are stable under small perturbations of structural parameters then the dynamical system is called structurally stable. This concept of stability is different from state stability, which is stability under the perturbations of state itself.

As a change of structural parameters, the equilibrium solution of the dynamical system, the pattern of equilibrium points and the trajectories of the system, bifurcate. Then a new stable pattern will emerge. This is called bifurcation. Structural change and self-organization are described by bifurcation of a suitably dynamical system.

Self-organization is explained as an emergence of a new equilibrium pattern of a dynamical system as a time and spatial order through a change of structural parameters.

Let us show the simple example of a one-dimensional dynamical system.

Let 0 be a state and structural stable equilibrium point in the following one-dimensional dynamical system. This is shown as follows.

$$0 \bullet \!-\!\!\leftarrow\!\!- \circ\, 1$$

Let the structural parameters change.

Then the following two patterns of the phase diagram show bifurcation patterns. The phase diagrams show the self-organization or structural change of the dynamical system.

$$0 \circ \!\rightarrow\!\!\!-\!\!\!-\!\!\!-\!\!\!- \bullet\, 1$$

$$0 \bullet \!-\!\!\leftarrow\!\!-\!\circ\!-\!\!\rightarrow\!\!- \bullet\, 1$$

Bifurcation analysis for structural change is a strong tool and has a broad application area. But it is almost incomplete for clarifying such systemic properties as hierarchy, emergent property and information processing.

Self-organization theory in physics argues structural change of the spatial pattern and time series of the concentration of molecules [Nicollis, 1977].

Let us consider the following chemical interactions.

$$A \rightarrow X,$$

$$B + X \rightarrow Y + D,$$

$$2X + Y \rightarrow 3X,$$

$$X \rightarrow E$$

Then kinetic equations of an intermediate product X, Y are shown as follows.

$$dX/dt = A - BX + X^2Y - X \qquad (2.1.1)$$

$$dY/dt = BX - X^2Y \qquad (2.1.2)$$

Where variables X and Y represent the concentration of chemical molecules.

The following Fig. 2.1 shows a convergence to the obvious equilibrium solution under such structural parameters as $A = 1$ and $B = 1$.

This model does not have spatial structural and no spatial patterns.

Instead a limit cycle as a pattern of time series occurs under the parameters $A = 1$ and $B = 3$. Figure 2.2 shows the limit cycle. Scatter diagram of the limit cycle is shown as follows. Bifurcation happens between $B = 1$ and $B = 3$ in this case.

We want to analyze structural change in such social sciences fields as social and economic systems. For this purpose it is necessary to construct a valid and reliable derivation method of a dynamical system, which represents our intended social and economic systems.

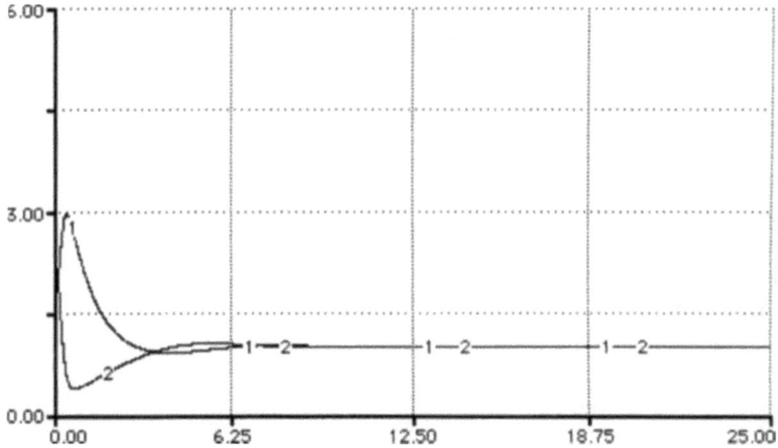

FIG. 2.1 Obvious Equilibrium

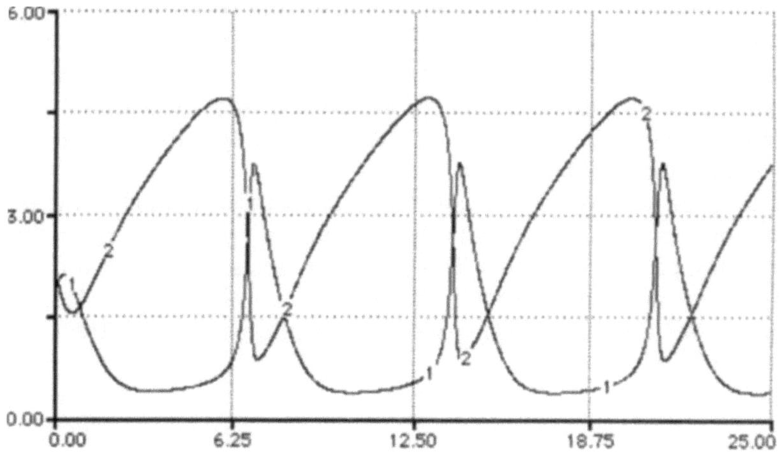

FIG. 2.2 Emergence of Limit Cycle

It is a hard question to answer. For example catastrophe theory is a theory of gradient types of dynamical systems and its bifurcation, which was developed by R. Thom [Thom, 1975; Zeeman, 1977].

The dynamics is shown as follows in general.

$$dx/dt = -\text{grad } V(x) \tag{2.1.3}$$

Many bifurcation models have developed in the area of social sciences as applications of catastrophe theory. The theory is excellent from a mathematical point

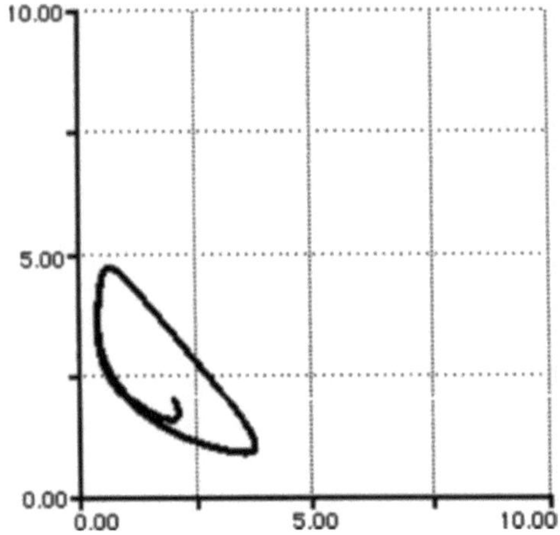

FIG. 2.3 Scatter Diagram of the Limit Cycle

of view, but unfortunately, most of the systems have no validity to social sciences, because it is difficult to find a gradient vector field in social science.

The validity of a model of a nonlinear dynamical system depends on the validity of the derivation method of the dynamical system. We want to derive nonlinear dynamical systems about decision making and its learning from the microscopic and agent based viewpoints. Then we analyze the structural change and self-organization of the system.

In this chapter we propose the bottom up derivation method of the dynamical system for social interaction with the learning of agents. We especially focus on the learning process of agents. This approach gives a theoretical foundation for agent based modeling as shown in the next chapter. We also focus on the bottom up reconstruction of state space in economic systems. We treat this topic in Chapters 5 and 6.

2.2 Game Theory and Evolutionary Games

Game theory is an important area in decision theory and economics [Shaun, 1995]. Game theory treats uncertainty (contingency) in decision making, where a choice of alternatives by an agent gives a landscape of evaluation for the mutual alternative selection of other agents. Game theory is a framework for analyzing this contingency. The uncertainty that we want to know under the multi-agent situation is a dynamic contingency in the process. Evolutionary game theory gives such a dynamic perspective.

At first non-cooperative game theory analyzes the rationality of decision making by using the concept of Nash equilibrium. However, we faced many difficulties when we expanded the concept of Nash equilibrium to one of a dynamic process.

The game theory of extensive form is used for analyzing the dynamic decision process of each agent in turn, which is expressed as a decision game tree. In this model the complexity of the game tree is very high.

A game tree search is used for board game analysis in artificial intelligence [Barr, 1986]. If the search is a vertical one then the state explosion problem will not happen. However a horizontal search is required to attain a sub game perfect Nash equilibrium in the tree analysis of the game theory of the extensive form.

There is little analysis of the computational complexity of the state explosion problem in game theory. The Traditional Nash equilibrium solution contains weak rationality in the extensive form of game theory. It is necessary to distinguish a weak and strong rationality. Then the concept of the Nash refinement is introduced and the sub game perfect Nash equilibrium is an example of Nash refinement.

We treat the concept of refinement on game theory in various ways, but such a refinement complicates the interpretation of the concept of a rational solution. If the purpose of rational decision theory exists in a formulation of normative decision making, then what kind of refinement will support the right rationality?

Our answer to the question is to shift our paradigm to learning dynamics. Of course, the problem "what kind of learning method is rational" is still left. Learning rationality also should connect to the concept of communicative rationality or procedural rationality [Habermas, 1989].

In addition we pay attention to the design of a learning system itself as a systems design. Some paradigm shift to learning is now occurring even in traditional economics [Fudenberg, 1998; Selten, 1991; Kandori, 1993; Nelson, 1982]. We not only study learning dynamics but also design an agent world under macro functional satisfying conditions. We also analyze structural change within an agent world.

In game theory we treat dynamical systems called replicator dynamics (RD) [Weibull, 1995; Andersen, 1996]. Replicator dynamics (RD) is one of the refinements of a Nash equilibrium in the context of repeated game theory. The dynamics of a population who select each alternative is induced by a random matching hypothesis in evolutionary game theory. The dynamics is called RD.

In this section we give a short survey of evolutionary game theory. Later we give a different way of deducing population dynamics, which represent the population ratio of alternative selections from the social learning point of view.

Let us consider the following decision making problem where an agent chooses an alternative from $\{a_1, a_2\}$. For example, {C, D} denotes the alternatives of the prisoner's dilemma game and {H, D} denote the alternatives of the hawk dove game. Let P denote the population ratio of the selection of alternative "a_1". The one dimensional dynamical system is normalized between [0,1] as follows.

$$dP(t)/dt = F(p(t)) \tag{2.2.1}$$

We induce this dynamics concretely from evolutionary game theory [Smith, 1982].

Let us consider E_1 and E_2 are payoff (profit) corresponding to each alternative respectively.

Then total average W is shown as $W = P_1E_1 + P_2E_2$.

M. Smith deduced a dynamical system called RD from an evolutionary point of view in biology. For the purpose M. Smith gives two hypotheses as follows.

(1) An agent repeats a random matching game and earns a payoff depending on his strategy. The accumulated payoff determines the adaptability of the agent.
(2) The increase of strategies in a population depends on the adaptability of the agent who selects the strategy

$$P(t + 1) = P(t)\{E_1/W\} \tag{2.2.2}$$

The following formula represents the hypothesis of multiplication depending on adaptability. The adaptability of the agent who selects "a_1" is shown as E_1/W.

Then the following formulae are deduced.

$$\Delta P = P(t + 1) - P(t) = P(t)\{E_1/W\} - P(t)\{W/W\}$$

$$= P(t)\{(E_1 - W)/W\} \tag{2.2.3}$$

$$dP(t)/dt = P(t)\{(E_1 - W)/W\} \tag{2.2.4}$$

RD is the dynamics of population ratio depending on a polymorphic interpretation of pure strategy in the game.

In relation to the Nash equilibrium, equilibrium points of RD, and asymptotic stability, the following properties are well known [Samuelson, 1997, pp 68–71].

(1) If (x,x) is a Nash equilibrium of a symmetric game then x is a stationary state of the replicator dynamics. If x is a stationary state, then (x,x) need not be a Nash equilibrium.
(2) If x is stable, then (x,x) is a Nash equilibrium. But there exist Nash equilibria (x,x) such that x is not stable.
(3) If x is asymptotically stable, then (x,x) is a perfect and isolated equilibrium. However neither perfect nor perfect and isolated equilibrium need be asymptotically stable.
(4) Every ESS (Evolutionary Stable Strategy) is asymptotically stable.

The asymptotic stable steady state gives a refinement of a Nash equilibrium solution.

Thus RD gives a solution to the problem of equilibrium selection.

In addition an asymptotically stable steady state of RD is a narrower solution than a perfect and isolated equilibrium.

RD is applied to a Comparative Institutional Analysis [Aoki, 2001]. A trajectory of RD is interpreted as an historical path. Stable steady states of RD are interpreted as institutions respectively. Where the plurality of stable steady states corresponds to the diversity of institutions. This indicates an application direction of a Nash refinement on RD.

This usage of RD is peculiar to economics, and different from biology. The stability concepts, which are used in RD, are state stable stability as in the sense of Lyapunov or in an asymptotically one. This is different from the more usual concept of structural stability.

However research interests do exist on structural stability in bifurcation theory. From this point of view we rather recognize realized institutional structural as a structural stable system itself, which is represented by the structural stable trajectory pattern of the dynamical system. The structural parameter of the dynamical system determines the variety of social or organizational structural as the structural stable equilibrium solution of the dynamical system. For this interpretation we introduced new types of dynamics deduced from the social learning point of view.

2.3 Social Learning Dynamics

We analyze the social learning process not from an evolutionary point of view but from a social learning point of view. In evolutionary game theory replicator dynamics (RD) is used for describing the dynamics of alternative selection. RD is usually derived from random matching a non-cooperative game that represents an evolutionary process. This is what we call an evolutionary interpretation of RD.

In the context of social science, the evolutionary interpretation is not satisfied in many cases. Instead, we introduce similar types of dynamics under the social learning interpretation. The new dynamics is called social learning dynamics [Deguchi, 1998, 2000a,b].

Let us consider ALT = {a, b} as a simple set of alternatives and an agent society that has a population of N. We introduce the following Markov process on the alternatives. Any player of the agent society changes his decision dynamically under the influence of an average payoff of each alternative.

For this purpose we will introduce the following transition probability.

$P_{ab}[1](t)$ denotes the transition probability to change the decision plan from alternative "a" to "b" at time "t" during a discrete time interval. $P_{ab}[h](t)$ denotes the transition probability to change a decision plan from "a" to "b" at time "t" during a small time interval "h".

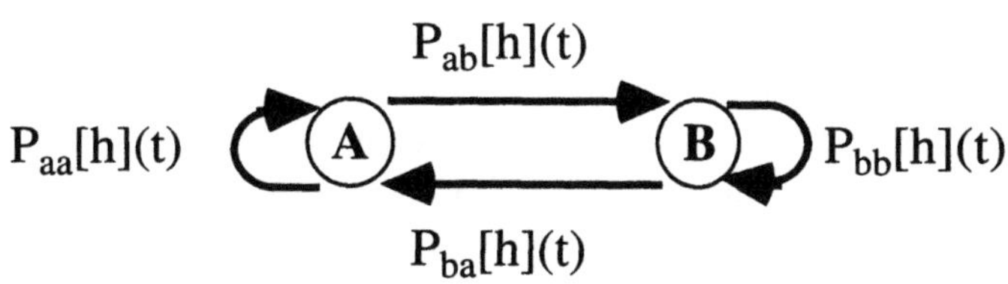

FIG. 2.4 Markov Learning Process of Alternatives

Let $P_{ab}[h](t) = h \times P_{ab}[1](t) = h \times P_b(t) \times E_b(t)/W(t)$,

$P_{ba}[h](t) = h \times P_{ba}[1](t) = h \times P_a(t) \times E_a(t)/W(t)$,

$P_{aa}[h](t) = 1 - h \times P_b(t) \times E_b(t)/W(t)$,

$P_{bb}[h](t) = 1 - h \times P_a(t) \times E_a(t)/W(t)$.

Where $E_a(t)$ and $E_b(t)$ denote average payoff (utility) of an agent who selects an alternative "a" and "b" respectively.

In addition, normalization condition of transition probability, $P_{cd}[h](t) + P_{cc}[h](t) = 1$ and $P_{dc}[h](t) + P_{dd}[h](t) = 1$ are required.

When "h" becomes 0 ($h \to 0$) then convergence conditions of transition probability are shown as $P_{cc}[h](t) \to 1$, $P_{dd}[h](t) \to 1$, $P_{dc}[h](t) \to 0$ and $P_{cd}[h](t) \to 0$.

Then we can derive a basic dynamical system, which represents the social learning process.

[Proposition 2.3.1] Social Learning Dynamics (SLD)

We can introduce the dynamics of decision probability, which is called social learning dynamics, as follows.

$$dP_a(t)/dt = P_a(t)\{E_a(t) - W(t)\}/W(t) \qquad (2.3.1)$$

$$dP_b(t)/dt = P_b(t)\{E_b(t) - W(t)\}/W(t) \qquad (2.3.2)$$

Proof:

$$P_a(t + h) = P_a(t) \times P_{aa}[h](t) + P_b(t) \times P_{ba}[h](t)$$

$$= P_a(t) + h \times P_a(t) \times P_b(t)\{E_a(t) - E_b(t)\}/W(t)$$

$$P_a[h](t) = P_a(t + h) - P_a(t) = h \times P_a(t) \times P_b(t)\{E_a(t) - E_b(t)\}/W(t)$$

$$dP_a(t)/dt = \lim_{h\to 0} \Delta P_a[h](t)/h = P_a(t) \times P_b(t)\{E_a(t) - E_b(t)\}/W(t)$$

$$= P_a(t)\{E_a(t) - W(t)\}/W(t)$$

Q.E.D.

The equation is diffeomorphic to $dP_c(t)/dt = P_c(t)\{E_c(t) - W(t)\}$. We now use this equation for bifurcation analysis instead of the original one.

There are two interpretations of the probability of this dynamic. The one is the probability of the selection from alternatives for any agent. In the first interpretation $P_a(t)$ represents the probability of selecting plan "a" for any agent. The other is the population ratio of selecting each alternative. In the second interpretation $P_a(t)$ denotes the population ratio of an agent society, who select plan "a".

It is also interpreted as the population ratio of the choice of alternatives. The two interpretations are consistent with each other.

In this model the average payoff is given in several ways. We do not use the average payoff deduced by the payoff matrix of a non-cooperative game. Of course random matching pairwise interactions of the game can give an average payoff. But we would like to introduce other methods to determine the average payoff in several ways. In the next chapter we introduce a mutual commitment mechanism, which determines the dynamic change of the average payoff under negative or positive social sanctions. We also introduce the bandwagon effect to give the average payoff.

Next we deduce the social learning dynamics for discrete alternatives.

[Proposition 2.3.2] Social Learning Dynamics for Discrete Alternatives
Let ALTS be a set of discrete alternatives.

$P_{ij}[1](t)$, $i\ j \in$ ALTS denote the transition probability from alternative "i" to "j" at time "t" for a discrete time interval. $P_j(t)$ denote the probability to select alternative "j" at time "t" for any agent, which also means a population ratio of selecting alternative "j".

Let $P_{ij}[1](t) = P_j(t) \times E_j(t)/W(t)$.

Where $W(t) = \sum\{P_j(t) \times E_j(t) \mid j \in \text{ALTS}\}$.

$E_j(t)$ denotes the expected average payoff (utility) of an agent by selecting alternative "j".

We assume the transition probability from "i" to "j" at time t during small time interval h as follows.

$$P_{ij}[h](t) = h \times P_{ij}[1](t) = h \times P_j(t) \times E_j(t)/W(t)$$

$$P_{ii}[h](t) = 1 - \sum\{P_{ij}[h](t) \mid j \in \text{ALTS}, j \neq i\}$$

As convergence conditions of transition probability we assume that

$$P_{ij}[h](t) \to 0 \text{ and } P_{ii}[h](t) \to 1 \text{ when } h \to 0.$$

Then we can deduce the differential equation of social learning dynamics as follows.

$$dP_i(t)/dt = P_i(t) \times \{E_i(t) - W(t)\}/W(t) \qquad (2.3.3)$$

Proof:

$P_i(t + h) = P_i(t) + \Delta P_i[h](t)$

$\Delta P_i[h](t) = \sum\{P_j(t) \times P_{ji}[h](t) \mid j \in \text{ALTS}, j \neq i\} - \sum\{P_i(t) \times P_{ij}[h](t) \mid j \in \text{ALTS}, j \neq i\}$

Then $dP_i(t)/dt$

$= \lim_{h\to 0} \Delta P_i[h](t)/h = \sum\{P_j(t) \times P_{ji}[h](t) - P_i(t) \times P_{ij}[h](t) \mid j \in \text{ALTS}, j \neq i\}/h$

$= \sum\{P_j(t) \times h \times P_i(t) \times E_i(t)/W(t)$
$\quad - P_i(t) \times h \times P_j(t) \times E_j(t)/W(t) \mid j \in \text{ALTS}, j \neq i\}/h$

$= P_i(t) \times \sum\{P_j(t) \times \{E_i(t) - E_j(t)\} \mid j \in \text{ALTS}, j \neq i\}/W(t)$

$= P_i(t) \times E_i(t) \times \sum\{P_j(t) \mid j \in \text{ALTS}, j \neq i\}/W(t)$
$\quad - P_i(t) \times \sum\{P_j(t) \times E_j(t) \mid j \in \text{ALTS}, j \neq i\}/W(t)$

$= P_i(t) \times E_i(t) \times \{1 - P_i(t)\} - P_i(t) \times \{W(t) - P_i(t) \times E_i(t)\}$

$= P_i(t) \times \{E_i(t) - W(t)\}/W(t)$ Q.E.D.

2.4 Asymmetrical Social Learning Dynamics

In the previous section we assumed learning interactions between agents in a single group. In this section we treat the learning interactions between two groups where an agent changes their attitude under the influence of agents from another group.

In this case we assume that there are two groups such as G_1 and G_2. An agent in group "G_1" has alternatives $\{a_1, a_2\}$ and an agent in group "G_2" has alternatives $\{b_1, b_2\}$.

Let $p(b_1)$ and $p(b_2)$ denote the population ratio of alternative "b_1" and "b_2" respectively. The average payoff of alternative $E(a_1)$ and $E(a_2)$ are affected by the population ratio of alternative "b_1" and "b_2" respectively.

Then $E[a_1] = U \times p[b_1] + H \times p[b_2]$.

The following table shows the effects of the other group. Then the average payoff for an agent in each group W and V are shown as follows.

$$W = q[a_1]E[a_1] + q[a_2]E[a_2] \tag{2.4.1}$$

$$V = p[b_1]E[b_1] + p[b_2]E[b_2] \tag{2.4.2}$$

Then we deduce the social learning dynamics of asymmetry interaction. We assume the following Markov process.

Transition probability p_{ij} and q_{ij} are defined as follows.

$$q_{12}[h](t) = h \times q_2 \times E[a_2]/W,\ q_{11}[h](t) = 1 - q_{12}[h](t) \tag{2.4.3}$$

$$q_{21}[h](t) = h \times q_1 \times E[a_1]/W,\ q_{22}[h](t) = 1 - q_{21}[h](t) \tag{2.4.4}$$

$$p_{12}[h](t) = h \times p_2 \times E[b_2]/V,\ p_{11}[h](t) = 1 - p_{12}[h](t) \tag{2.4.5}$$

$$p_{21}[h](t) = h \times p_1 \times E[b_1]/V,\ p_{22}[h](t) = 1 - p_{21}[h](t) \tag{2.4.6}$$

TABLE 2.4.1 Interaction Coefficient from Other Groups

Average Payoff	Effect of b_1	Effect of b_2
$E[a_1]$	$U \times p[b_1]$	$H \times p[b_2]$
$E[a_2]$	$T \times p[b_1]$	$V \times p[b_2]$
Average Payoff	Effect of a_1	Effect of a_2
$E[b_1]$	$R \times q[a_1]$	$S \times q[a_2]$
$E[b_2]$	$G \times q[a_1]$	$Q \times q[a_2]$

TABLE 2.4.2 Asymmetry Payoff Game Between Two Groups

B\A	a_1	a_2
b_1	(R, U)	(S, T)
b_2	(G, H)	(Q, V)

FIG. 2.5 Asymmetry Markov Learning Process of Alternativesa

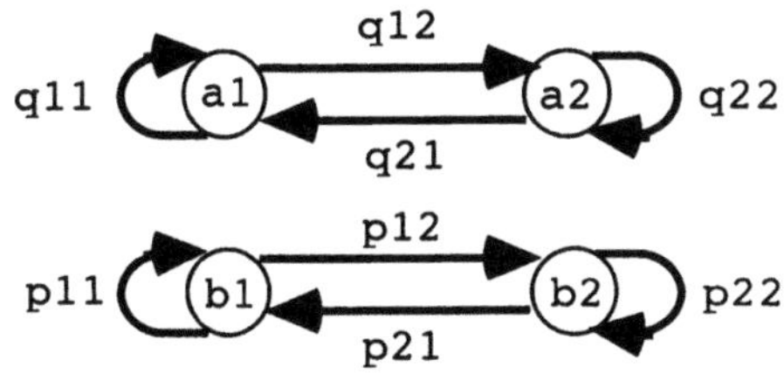

Then the social learning dynamics are derived as follows.

$$\begin{aligned}\Delta p[b_1][h](t) &= p[b_1](t+h) - p[b_1](t)\\ &= p[b_1](t) \times p_{11}[h](t) + p[b_2](t) \times p_{21}[h](t) - p[b_1](t)\\ &= p[b_1](t) \times (1 - h \times p[b_2] \times E[b_2]/V)\\ &\quad + h \times p[b_2](t) \times p[b_1](t) \times E[b_1]/V - p[b_1](t)\\ &= h \times p[b_2](t) \times p[b_1](t) \times \{E[b_1] - E[b_2]\}/V \end{aligned} \tag{2.4.7}$$

$$\begin{aligned} dp[b_1](t)/dt &= \lim_{h\to 0} \Delta p[b_1][h](t)/h\\ &= p[b_2](t) \times p[b_1](t) \times \{E[b_1](t) - E[b_2](t)\}/V(t)\\ &= p[b_1](t) \times \{E[b_1](t) - V(t)\}/V(t) \end{aligned} \tag{2.4.8}$$

$$dp[b_1]/dt = \{P[b_1](E[b_1] - V)\}/V \tag{2.4.9}$$

$$dq[a_1]/dt = \{q[a_1](E[a_1] - W)\}/W \tag{2.4.10}$$

In the case of a random matching asymmetry game in an evolutionary game, we can deduce the interaction coefficient table from the following asymmetry payoff matrix.

We can apply asymmetrical social learning dynamics to many areas. For example consumer and seller interact with each other in the real economy. We apply this SLD in Chapter 9 for analyzing industrial structurals.

Chapter 3
A Social and Organizational Learning Model of Decision Making
—Application of SLD and Agent Based Simulation—

Purpose of This Chapter

In this chapter learning, evolution and the structural change of multi agent systems are investigated using actual examples of social and economic systems.

First we consider the prisoner's dilemma problem and the structural change of the game and institutional design by using social learning dynamics (SLD). We introduce the concept of indirect control of an agent society, where we control the boundary conditions of an agent society to satisfy the macroscopic function.

We especially pay attention to structural change by bifurcation of the dynamical system. A norm game and a Meta norm game by R. Axelrod are analyzed by SLD. We shall introduce central and mutual commitment models for social learning dynamics. We also look at more complex decision models of agent society, which include complex mutual commitment.

3.1 Indirect Control

We will start with the ever-popular model of the prisoner's dilemma, but we will not focus on the usual state stability and historical path, but on the structural stability of dynamics of the model.

We introduce an agent society that consists of a random matching model of the prisoner's dilemma. In the society agents make random matching pairs and play two person prisoner's dilemma as shown in the following table. Under the random matching assumption an agent can not appoint an opposite player.

We assume $R = 2, S = 0.5, T = 3, Q = 1$ in the following example. We assume that $T > R > Q > S$ and $R > (T + S)/2$ from the condition of prisoner's Dilemma. Let P be the ratio of the population of agents who select the cooperative alternative (C).

[Average Payoff]

$E(C,C) = R, E(D,D) = Q, E(C,D) = S, E(D,C) = T$

$E[C] = PE(C,C) + (1 - P)E(C,D) = PR + (1 - P)S$

TABLE 3.1 Payoff Matrix of Prisoner's Dilemma

	C	D
C	(R,R)	(S, T)
D	(T, S)	(Q,Q)

$$E[D] = PE(D,C) + (1 - P)E(D,D) = (1 - P)Q + PT$$
$$W = PE[C] + (1 - P)E[D] = PR + (1 - P)S + (1 - P)Q + PT$$

[Dynamics]

$$\begin{aligned} dP/dt &= P(E[C] - W) = P\{E[C] - PE[C] - (1 - P)E[D]\} \\ &= P\{(1 - P)(E[C] - E[D])\} \\ &= P(1 - P)\{P(R + Q - S - T) + S - Q\} = 0, \text{ where } 0 \leq P \leq 1 \text{ hold.} \end{aligned}$$

$dP/dt = 0$ means steady states of the replicator dynamics. Then (1) $p = 1$, (2) $p = 0$ and (3) $p = (Q - S)/(R + Q - S - T)$ are possible steady states.

Where $0 < P = (Q - S)/(R + Q - S - T) < 1$ is required. From the assumption $P = (Q - S)/(R + Q - S - T) > 1$ holds.

Thus there exist only two possible steady states such as $p = 1$ or $p = 0$.

$dP/dt < 0$ also holds. Thus only $P = 0$ is stable, which means, as the game theory state, D is a dominant solution with these dynamics.

Next we try to formulate an indirect control by varying the boundary conditions of social learning dynamics. We introduce a decision making model incorporating the risk of the prisoner's dilemma model.

[Example 3.1] Indirect Control of PD Type Society

We assume the macro agent such as a government or a manager who is monitoring agent activities with a certain probability.

Let β be a probability of the monitoring. Let "a" be a compensation for a betrayed agent and "b" be a negative sanction for a betrayer. The payoff matrix is shown as follows.

[Payoff]

$$E[\beta](C,C) = R,\ E[\beta](C,D) = S + a,$$
$$E[\beta](D,C) = T - b,\ E[\beta](D,D) = Q$$

[Average Payoff]

$$\begin{aligned} E[\beta,C] &= \beta\{PE[\beta](C,C) + (1 - P)E[\beta](C,D)\} \\ &= \beta\{PR + (1 - P)(S + a)\} \\ E[\beta,D] &= \beta\{PE[\beta](D,C) + (1 - P)E[\beta](D,D)\} \\ &= \beta\{P(T - b) + (1 - P)Q \end{aligned}$$

$$E[1 - \beta,C] = (1 - \beta)E[C] = (1 - \beta)\{PR + (1 - P)S\}$$
$$E[1 - \beta,D] = (1 - \beta)E[D] = (1 - \beta)\{(1 - P)Q + PT\}$$

TABLE 3.2 Payoff Matrix of Prisoner's Dilemma with a Monitoring Risk

	C	D
β	C (R,R)	(S + a, T − b)
	D (T − b, S + a)	(Q,Q)
$1-\beta$	C (R,R)	(S, T)
	D (T, S)	(Q,Q)

$$
\begin{aligned}
E[C] &= E[\beta,C] + E[1-\beta,C] \\
&= \beta\{PR + (1-P)(S+a)\} + (1-\beta)\{PR + (1-P)S\} \\
E[D] &= E[\beta,D] + E[1-\beta,D] \\
&= \beta\{P(T-b) + (1-P)Q\} + (1-\beta)\{(1-P)Q + PT\} \\
W &= PE[C] + (1-P)E[D]
\end{aligned}
$$

[Dynamics]

$$
\begin{aligned}
dP/dt &= P\{E[C] - W\} = P(1-P)\{E[C] - E[D]\} \\
&= P(1-P)\{\beta\{PR + (1-P)(S+a)\} + (1-\beta)\{PR + (1-P)S\} \\
&\quad - \beta\{P(T-b) + (1-P)Q\} - (1-\beta)\{(1-P)Q + PT\}\} \\
&= P(1-P)[(\beta a + S - Q) + P\{\beta(b-a) + R - S + Q - T\}]
\end{aligned}
$$

Where $dP/dt = 0$ means steady states of the social learning dynamics. Then $p = 1$, $p = 0$ and $P^* = (Q - \beta a - S)/\{\beta(b-a) + R - S + Q - T\}$ are possible steady states.

The followings are the examples of structural parameters for an indirect control.

(1) $\beta = 0.8$, $a = 1$, $b = 1.5$ 0 ○—>—• 1
(2) $\beta = 0.6$, $a = 1$, $b = 1.5$ 0 ○—>•<—○ 1
(3) $\beta = 0.4$, $a = 1$, $b = 1.5$ 0 •—<—○ 1

Where "•" means stable state and "○" means unstable state and (β, a, b) become structural parameters of this dynamical system.

[Indirect Control by Government]

The purpose is to maximize the total social payoff. To achieve our aim we use parameters (β, a, b) as a means to indirectly lead the agent activities towards cooperative ones.

In the previous case a government can vary the parameter β under the conditions of $a = 1, b = 1.5$ to make agent activities become cooperative. $\beta = 0.8$ is enough to change most of the autonomous and rational activities of agents into cooperative ones.

We will show another example with a structural parameter $\beta = 0.4$, $a = 1$ and $b = 3$. Where the stability of the states are shown as follows.

(4) $\beta = 0.4, a = 1, b = 3$ 0 •—<—○—>—• 1

General Condition for $0 < P^* < 1$ is shown as follows.

$$P^* = (Q - S - \beta a)/\{\beta(b - a) + R + Q - (S + T)\}$$

(A) $\beta(b - a) + R + Q - (S + T) > 0$
P^* is Unstable under the following conditions.

$$\{(T - R)/b\} < \beta < \{(Q - S)/a\}$$

$$(Q - S) < a, (T - R)/b < (Q - S)/a$$

The equilibrium states are shown as follows. Where white circles indicate unstable equilibrium states and black circles indicate stable equilibrium states.

0 •—<—○—>—• 1

(B) $\beta(b - a) + R + Q - (S + T) < 0$
P^* is Stable under the following conditions.

$$\{(Q - S)/a\} < \beta < \{(T - R)/b\}$$

$$(T - R) < b, (Q - S)/a < (T - R)/b$$

The equilibrium states are shown as follows. In this case there is only one stable state.

0 ○—>—•—<—○ 1

3.2 Norm Formation, Mutual Commitment and Structural Change

Norm formation is one of the most basic processes of social interaction. To clarify this process, we introduce mutual commitment and meta commitment among agents. R. Axelrod has analyzed the commitment structure of norm formation and the collapsing process of the norm with his norm and meta norm games [Axelrod, 1997, Ch.3]. He analyzes the games with genetic algorithms.

In this section we generalize the games as mutual commitment processes of agents [Deguchi, 2000a,b]. We also introduce the commitment by central authorities such as a government that enforces legal policies. We deal with the model from a theoretical point of view. For the purpose of describing a social learning, we introduce social learning dynamics (SLD). We have already deduced SLD from a stochastic learning process in the previous chapter.

Now we reformulate the norm game as a coupling model between the social learning dynamics of alternatives and the dynamics of a normative attitude. We also extend this model and introduce several types of centralized and decentralized commitment mechanisms on agent societies.

In this chapter we focus on the concept of norm as a decentralized indirect regulation on a society. For this purpose, we analyze the mutual commitment

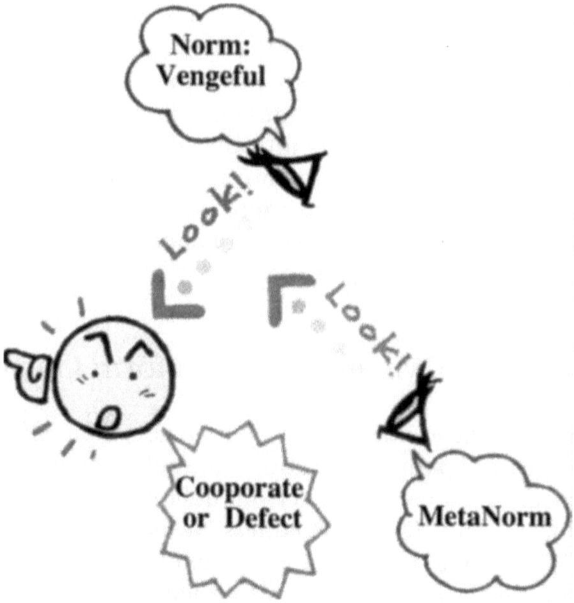

FIG. 3.1 Mutual Reference and Commitment

and the meta commitment processes among agents as shown in Fig. 3.1. The concept of norm is analyzed in the multidisciplinary area of social science. It is difficult to give it a suitable definition but R. Axelrod has introduced the essential definition. He defines norm and meta norm games and introduces the mutual and hierarchical commitment structures of these games. He explains how norms are collapsed and maintained in the games.

We reformulate his norm and meta norm games using social learning dynamics. Social learning dynamics (SLD) was introduced in Chapter 3, which describe the social learning process. Axelrod introduces his norm game as an extension of n-person Prisoner's Dilemma. He adds a new feature of normative attitude to the dilemma game, which is to punish defectors under the mutual monitoring and commitment processes.

In the norm game, the strategies of players consist of two dimensions. He introduces two set of alternatives, {Boldness, No-Boldness} and {Vengefulness, No-Vengefulness}. The first represents if a player is bold enough to defect or not. In other words, it is an attitude to follow, or not to follow, the norm. The second represents if a player is vengeful enough to punish the player who does not follow the norm. It is an indirect regulation of the norm through mutual commitment to the other person's attitude. He analyzes the norm game by genetic algorithm and shows how norms collapse because of the retrogression of normative commitment.

He also shows how to prevent the retrogression of normative commitment and how to maintain the norm. He introduces the concept of meta norm as a mutual meta commitment for normative commitment. It means that a player has enough vengefulness to punish the player who does not execute his normative commitment. If the meta norm mechanism is introduced in a society then normative commitment does not retrogress and the norm does not collapse.

We reformulate his norm game and meta norm game with social learning dynamics. For this purpose, we introduce the concept of decentralized indirect regulation or indirect control of agent society.

3.3 Social Learning Dynamics of the Norm Game

In the norm game, if a player does defect then he gets a payoff of T = 3. Pd and Pc denote the normalized population ratio of the selection of Boldness and No-Boldness respectively. Pc = 1 − Pd holds from definition. In this section we used "C" for "No-Boldness" and "D" for "Boldness" respectively. PV and Pnv denote the normalized population ratio of the selection of Vengefulness and No-Vengefulness respectively. Pnv = 1 − PV also holds.

As pointed out earlier, "Vengefulness" represents the normative attitude under the mutual commitment process that is illustrated as follows (Fig. 3.1).

For formulating the social learning dynamics of the norm game we introduce an average payoff structure as follows according to Axelrod's game structure. Then we can introduce SLD for the norm game as follows. Where we assume a < d < b < c. Let a = 1, b = 3, c = 9, d = 2 in the following simulation.

$$dPc/dt = Pc \times (1 - Pc) \times (EC - ED) \quad (3.3.1)$$

$$dPV/dt = PV \times (1 - PV) \times (EV - Env) \quad (3.3.2)$$

$EC = -a \times (1 - Pc)$
$ED = b \times Pc - c \times PV$
$Env = 0$
$EV = -d \times (1 - Pc)$
$pd = 1 - Pc$
$Pnv = 1 - PV$

TABLE 3.3 Average Payoff of Underlying and Commitment of Alternatives

Average Payoff	Interaction Factor	Commitment Factor
E[C]	−a × Pd	0
E[D]	b × Pc	−c × PV
E[V]	0	−d × Pd
E[NV]	0	0

[Proposition 3.3.1]
Pc = 1 is a stable steady state of the formula (3.3.1) if and only if

$PV > \{(b - a)Pc + a\}/c$

Proof:
Pc = 1 is a steady state of the formula (3.3.1). The stability depends on the sign of EC – ED.

$$dPc/dt > 0 \Leftrightarrow EC - ED > 0 \Leftrightarrow -a \times (1 - Pc) > b \times Pc - c \times PV$$
$$\Leftrightarrow PV > \{(b - a)Pc + a\}/c \qquad \text{Q.E.D.}$$

We add the perturbation factor to this model. The result shows the collapsing process of the norm (Fig. 3.2) and it corresponds to Axelrod's simulation by using a genetic algorithm (GA) under the Complex Adaptive System (CAS) paradigm.

[Proposition 3.3.2]
(1) Let $PV > \{(b - a)Pc + a\}/c$. Then $|dPV/dt| \leq d \times \varepsilon/4$, where $\varepsilon = 1 - Pc$ is small perturbation.
(2) Let $PV < \{(b - a)Pc + a\}/c$. Then $|dPV/dt| \leq d \times \alpha/4$, where $\alpha = 1 - Pc$ become 1 as a result.

Proof:
(1) $PV > \{(b - a)Pc + a\}/c$ means $dPc/dt > 0$ from Proposition 3.3.1. Then Pc becomes 1 with small perturbation ε.
$0 \leq PV \times (1 - PV) \leq 1/4$, $EV = -d \times (1 - Pc) = -d\varepsilon$ and $Env = 0$ hold.
Then $|dPV/dt| = |PV \times (1 - PV) \times (EV - Env)| \leq d \times \varepsilon/4$.

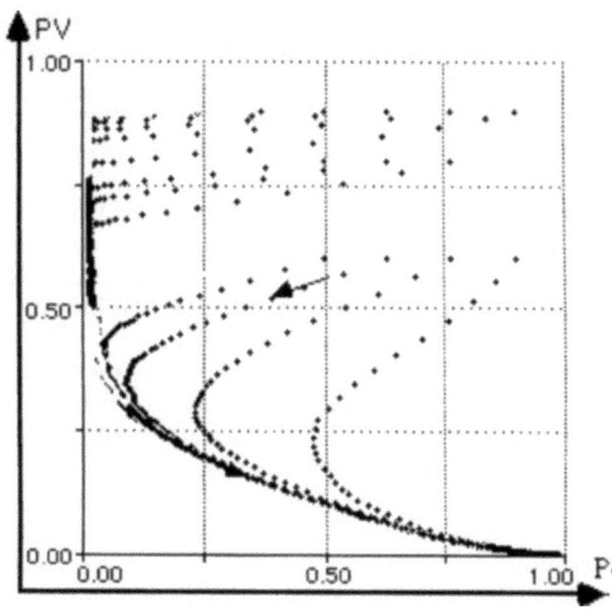

FIG. 3.2 Norm Collapse

(2) $PV < \{(b-a)Pc + a\}/c$ means $dPc/dt < 0$ from Proposition 3.3.1. Then Pc becomes 0.

Then $|dPV/dt| = |PV \times (1 - PV) \times (EV - Env)| \leq d \times \alpha/4$ Q.E.D.

The Proposition 3.3.2 (1) asserts that the decreasing rate of PV is small and in proportion to ε.

The Proposition 3.3.2 (2) asserts that the decreasing rate of PV becomes large in proportion to α.

Next we show the role of a small group acting as the salt of the earth or catalyst who prevent the collapse of the norm by increasing the average payoff when $PV > \{(b - a)Pc + a\}/c$ holds.

[Proposition 3.3.3]
Let $EV = -d \times \varepsilon + \Delta$. Then the collapse of norm is prevented if $\Delta > d \times \varepsilon$ hold.

Proof:

$EV = -d \times (1 - Pc) + \Delta = -d\varepsilon + \Delta$ and $Env = 0$ hold.

Then $EV - Env > 0 \Leftrightarrow -d\varepsilon + \Delta > 0 \Leftrightarrow \Delta > d\varepsilon$ hold. Q.E.D.

The Proposition 3.3.3 asserts that the decreasing rate of PV is prevented by a small compensation made by the central authority or the efforts of the salt of the earth.

The norm game provides a new conceptual model of the norm. The GA simulation provides experimental discovery such as the collapsing process of the norm. The GA is a complex and complicated algorithm. It is difficult to extract systemic properties from the simulation. It is also difficult to extend the model to a more general centralized and decentralized regulation processes of an agent society under mutual commitments.

We reformulate the mechanism with the population dynamics called SLD. Once we give the mathematical formulation, it is easy to analyze and extend the mechanism.

The model consists of two 1 dimensional SLD and its weak coupling. The population ratio of Vengefulness (V) becomes a bifurcation parameter for the dynamics of underlying alternatives {C,D}. If PV is large enough then Pc increases. If V becomes small then the system bifurcates and Pd increases. The bifurcation structure is shown in Fig. 3.3.

R. Axelrod also introduced the concept of meta norm that is included in the concept of meta commitment among agents. It is illustrated in Fig. 3.4. In this figure, the thick arrows indicate normative commitments and the thin arrows indicate meta normative commitments.

We introduce the average payoff table of commitment and meta commitment of alternatives as follows. Then we can introduce SLD for meta norm game as follows. Where we assume $a < d < b < c = e$. Let $a = 1, b = 3, c = 9, d = 2, e = 9$ in the following simulation. We also introduce fluctuation in the simulation.

FIG. 3.3 Bifurcation of Pc Dynamics

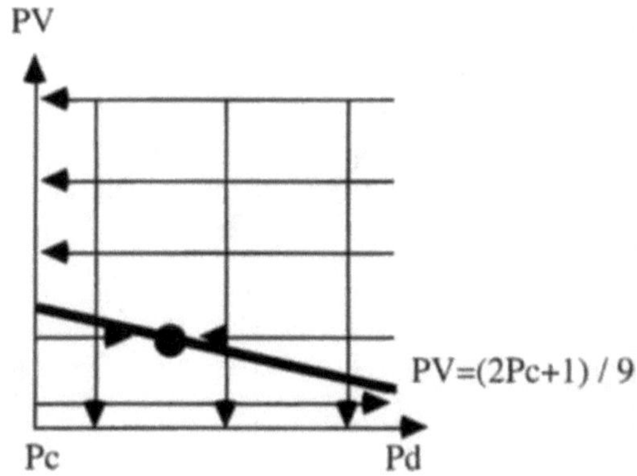

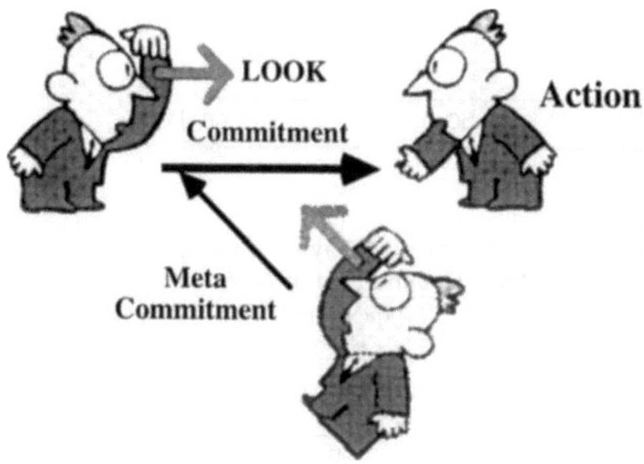

FIG. 3.4 Commitment and Meta Commitment

$$dPc/dt = Pc \times (1 - Pc) \times (EC - ED) \tag{3.3.3}$$

$$dPV/dt = PV \times (1 - PV) \times (EV - Env) \tag{3.3.4}$$

$$dPMV/dt = PMV \times (1 - PMV) \times (EMV - EMnv) \tag{3.3.5}$$

$EC = -a \times (1 - Pc)$, $ED = b \times Pc - c \times PV$
$EMnv = 0$, $EMV = -d \times (1 - PV)$
$Env = -e \times PMV$, $EV = -d \times (1 - Pc)$
$Pd = 1 - Pc$, $Pnv = 1 - PV$
$PMnv = 1 - PMV$

Figure 3.5 shows the result of the simulation of the meta norm.

We have the same question of whether the meta norm also collapses or not. This is also a question for a second-order social dilemma. The following proposition answers these questions.

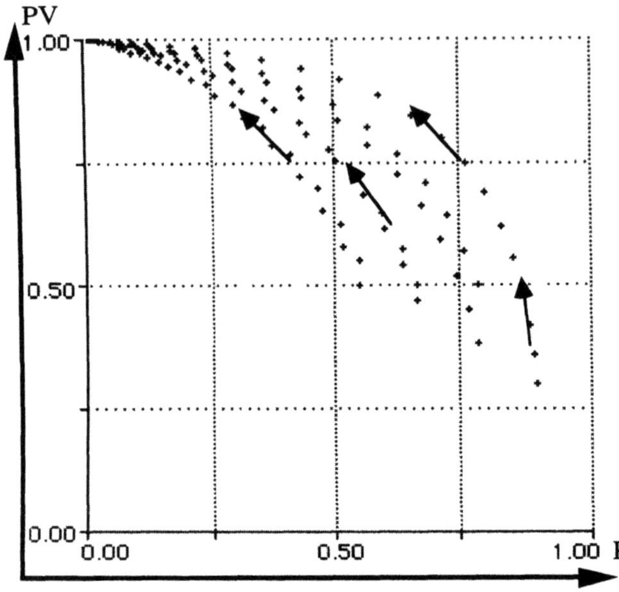

FIG. 3.5 SLD for Meta Norm

TABLE 3.4 Average Payoff of Commitment and Meta Commitment of Alternatives

Average Payoff	Interaction Factor	Coupling Factor
E[V]	0	–d × Pd
E[NV]	0	–e × PMV
E[MV]	0	–d × Pnv
E[Mnv]	0	0

[Proposition 3.3.4]
Let ε and ε′ be positive perturbation around Pc = 1 and positive perturbation around PMV = 0 respectively. If ε and ε′ are the same scale of positive perturbation and c > d then PV = 1 is stable.

Proof:
We show that dPV/dt > 0. dPV/dt = PV × (1 – PV) × (EV – Env).

Then dPV/dt > 0 ⇔ EV > Env ⇔ –d × (1 – Pc) > –c × PMV holds. From the assumptions –d × (1 – Pc) > –c × PMV ⇔ –dε > –c × ε′ holds and ε and ε′ are same scale of perturbation.

Thus we can assume that ε = ε′. Then c > d is the condition for dPV/dt > 0.

Q.E.D.

In the previous example c = 9, d = 2 satisfy the condition.

3.4 Centralized and Hierarchical Commitment and Support

3.4.1 Type of Commitments

Once we formulate the mathematical model of the norm and analyze its bifurcation mechanism, we can extend the model to the more general processes of mutual reference and commitments. We add a centralized commitment structure in the society. We also introduce a positive sanction mechanism as a commitment process such as a supporting attitude.

In the previous sections we have analyzed the mutual reference and commitment processes. We call the mechanism decentralized commitment, which effects an indirect regulation of agent societies. The norm and meta norm affect the underlying alternatives as a decentralized indirect regulation. Now we shall try to design another indirect regulation mechanism for agent societies.

In this section we assume that there is a central authority such as government in the society and introduce centralized commitment processes. For example monitoring and a fine for a parking violation is a typical centralized commitment for underlying actions, which is shown in Fig. 3.6. We also pay attention to the centralized commitment for mutual commitment processes and meta commitment processes, which is shown in Fig. 3.7.

We also introduce a positive sanction as a commitment to the model. In the norm and meta norm games, the sanction is a negative penalty. Instead, we can use a positive sanction such as supporting commitment instead of a negative sanction.

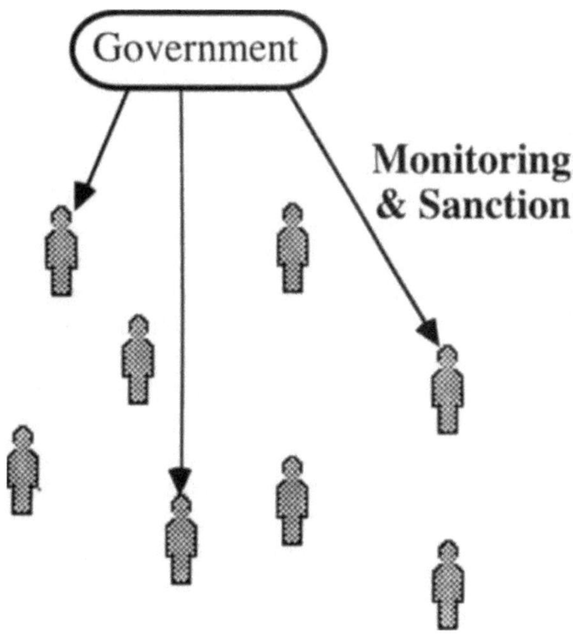

FIG. 3.6 Centralized Monitoring and Sanction for Underlying Actions

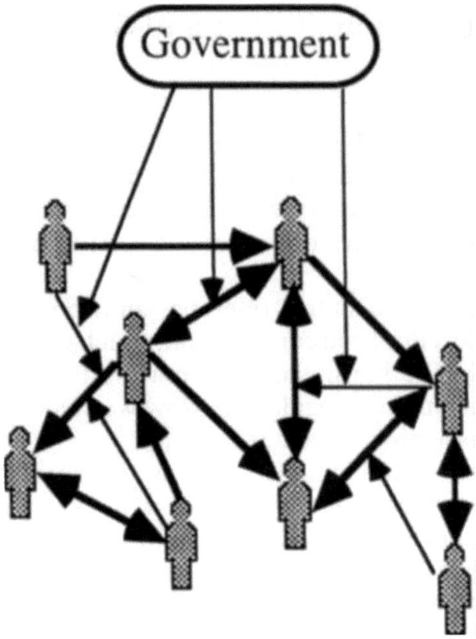

FIG. 3.7 Centralized Monitoring and Sanction for Mutual Commitments

	A	B
A	(a, a)	(−b, 0)
B	(0, −b)	(−a, −a)

3.4.2 Educational Effect and Supporting Commitment

To clarify our new concepts, we constructed the model of educational effect and its support by government as follows. We introduce the underlying alternatives "A" and "B". In this case we assume that "ASup" and "BSup" are alternatives that mean supporting the underlying alternatives respectively. Alternatives "AKnow" and "Bknow" mean the belief that "A" is correct and "B" is correct respectively.

In the model we assume that "A" is a desirable attitude in the long run for society and "B" is a conventional attitude. We assume the dilemma situation shown in the Table 3.5 under the interpretation of random matching game. When we consider the basic interaction level under the social learning interpretation, we get the interaction table of average payoff as a part of Table 3.6. Until this stage we can use the both interpretation. Next we add mutual and centralized commitments. We assume that there is political support for the "Aknow". The initial conditions for the simulation assume that old beliefs such as "PbKnow" and "PBSup" are dominant. But the educational support changes the situation drastically. As the right knowledge spreads, agents who support the right manner increase, then the population of "A" spreads.

TABLE 3.6 Average Payoff for Alternatives of Supporting Commitments

Average Payoff	Interaction with Different Type	Interaction with Same Type	Coupling and Political Support
E[A]	−b × Pb	a × Pa	c × PASup
E[B]	0	−a × Pb	0
E[ASup]	0	d × Pa	e × PaKnow
E[BSup]	0	d × Pb	0
E[AKnow]	0	0	polsupA = 1
E[BKnow]	0	0	0

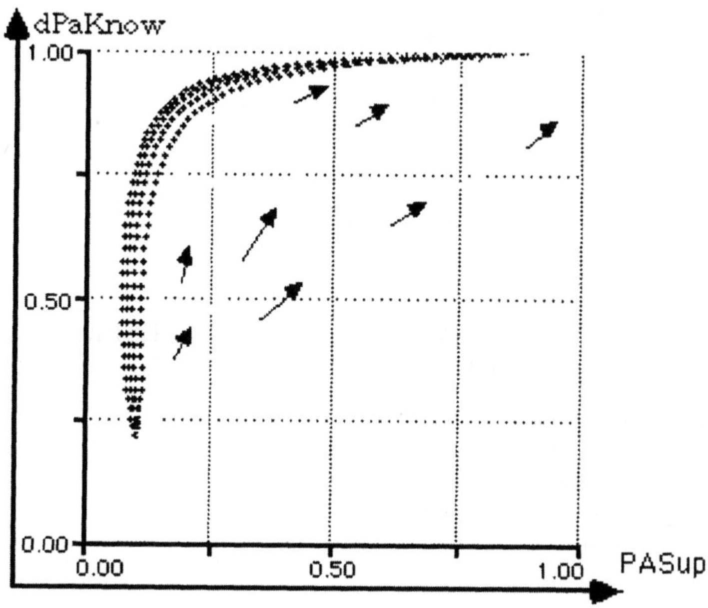

FIG. 3.8 The Spread of Correct Knowledge

Furthermore, we consider the following positive sanction as the commitment shown by "ASup" and "BSup". We also assume meta commitment from the correct knowledge shown by "AKnow" and "BKnow". Where "a", "b", "d", and "e" mean basic payoff of the interaction, negative sanction from conservatives, bandwagon effects, and positive support by correct knowledge respectively.

At the first stage we are bound by convention. How can we do away with conventionalities? We are too conventional at the first stage because of the sanctions from conservatives and the bandwagon effects. We assume a < b. Let a = 2, b = 5, c = 3, d = 1, and e = 2 in the following simulation. Then we can introduce SLD for the educational effect and supporting commitment as follows. We also introduce fluctuation in the simulation.

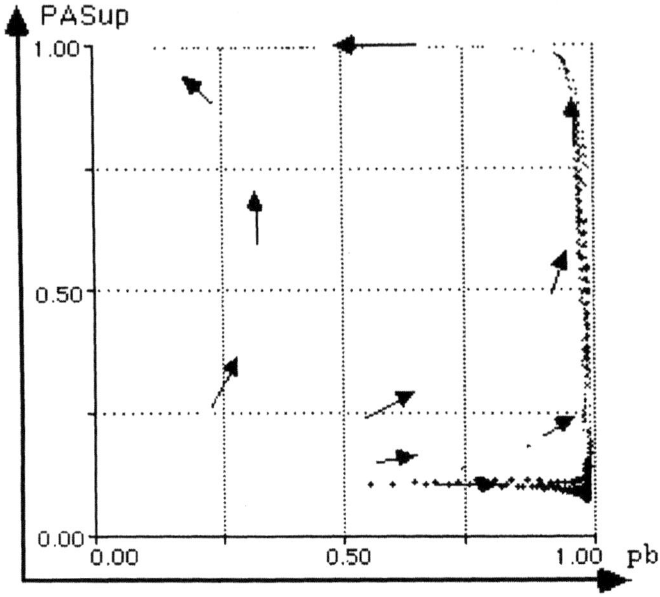

FIG. 3.9 The Spread of the Right Attitude

$$dPa/dt = Pa \times (1 - Pa) \times (EA - EB)$$
$$dPaKnow/dt = PaKnow \times (1 - PaKnow) \times (EAKnow - EBKnow)$$
$$dPASup/dt = PASup \times (1 - PASup) \times (EASup - EBSup)$$

$EA = -b(1 - Pa) + aPa + c \times PASup,$
$EB = 0 \times Pa - a(1 - Pa) + 0 \times (1 - PASup)$
$EASup = Pa + e \times PaKnow,$
$EBSup = (1 - Pa) + 0 \times (1 - PaKnow)$
$EAKnow = polsupA,$
$EBKnow = 0,$
$polsupA = 1$
$Pb = 1 - Pa,$
$PBSup = 1 - PASup,$
$PbKnow = 1 - PaKnow$

[Proposition 3.4.1]
Let $PASup = 0$. This means that there is no supportive commitment for attitude "A".

(1) If $a < b$ then $Pb = 1$ is stable, i.e., an existing conventional attitude is not overthrown.
(2) If $a > 0$ then $Pa = 1$ is stable, i.e., an existing desirable attitude is maintained.

Proof:
(1) We show that $dPa/dt < 0 \Leftrightarrow EA - EB < 0 \Leftrightarrow -b \times (1 - Pa) + a \times Pa + c \times PASup + a \times (1 - Pa) < 0$

$\Leftrightarrow$ a − b + c × PASup + b × Pa < 0
$\Leftrightarrow$ a + c × PASup < b − b × Pa
If PASup = 0, Pb = 1, and a < b then dPa/dt < 0 holds.
(2) dPa/dt > 0 $\Leftrightarrow$ a + c × PASup > b − b × Pa.
Let Pa = 1 and a > 0 then dPa/dt > 0 holds even if PASup = 0. Q.E.D.

We are interested in the overthrowing process of the existing conventional attitude.

The proof shows that if PASup is large enough and a + c × PASup > b then dPa/dt > 0 holds for any Pa > 0.

[Proposition 3.4.2]
Let PaKnow = 1 then e > d is the condition of spreading supportive commitment of attitude "A" for any Pa.

Proof:
We show that dPASup/dt > 0.
dPASup/dt = PASup(1 − PASup)(EASup − EBSup)
dPASup/dt > 0 $\Leftrightarrow$ EASup − EBSup > 0 $\Leftrightarrow$ e × PaKnow + d × Pa − d × (1 − Pa) > 0 $\Leftrightarrow$ e × PaKnow + d × Pa − d + d × Pa > 0 $\Leftrightarrow$ e × PaKnow > d − 2d × Pa
We assume that PaKnow = 1 then e × PaKnow > d − 2d × Pa $\Leftrightarrow$ e > d − 2d × Pa.
Then e > d is the condition of dPASup/dt > 0 for any Pa. Q.E.D.

We can show a real life example of this model. The United Nations has put a great deal of effort into the furtherance of birth control, but a policy of centralized direct sanctions has failed. People do not want to change their beliefs. Now they say that the spread of right knowledge and its support for birth control are the only way to change people's behaviors.

3.5 The Historical Change of Social Commitment Structure

Once we recognize hierarchical mutual commitment among agents, we can easily explain norm formation and other social orders and changes. Furthermore, we can characterize historical change of the commitment structure in our society. In pre-modern society, decentralized commitment such as meta norm is a typical way of mutual commitment where the sanction is a negative one such as punishment.

Central authority might support the meta norm process, which is shown in Fig. 3.10.

In pre-modern society, it is difficult to change social order quickly, in contrast modern society is highly fluid and varied and requires dynamic change for its economical development. In society, a new way of forming the social norm is required. Under the nation state, the old way of mutual sanctions became out-of-date. Two types of inducing commitment have been introduced by the nation state. One is the rule of law and the other is education by the mass media and schools.

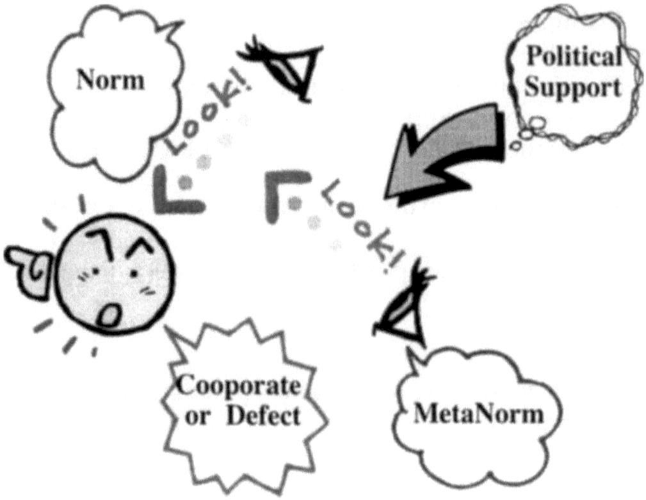

FIG. 3.10 Indirect Control on Pre-Modern Society

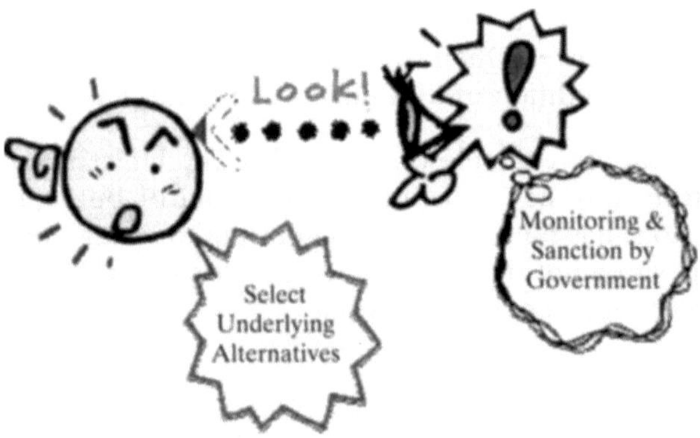

FIG. 3.11 Monitoring and Punishment by Law

If someone breaks the law then they would be judged and punished according to law under the social monitoring, which is shown in Fig. 3.11.

Another way of social commitment has developed in modern society. It consists of a positive commitment such as support for underlying attitudes, meta commitment depending on knowledge and information and educational support of knowledge. In modern society knowledge formation is supported or affected by education or the mass media. The process is shown in Fig. 3.12.

Of course this classification is a representative one in the period. There are different types of commitment structures in any period. In modern society, imposed evaluation of knowledge by education and the mass media is dominant. A new type of mutual commitment structure is rapidly developing through the Internet and IT mediated world.

For evolving the evaluation landscape of underlying activities dynamically, we

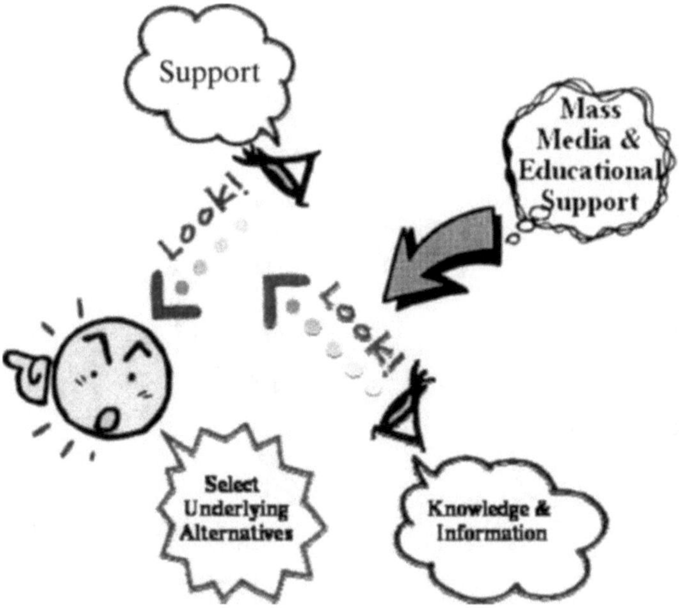

FIG. 3.12 Meta Commitment Through Knowledge and Information and Its Support

have to learn knowledge about the situation. Knowledge creates our evaluation landscape, which dominates our supporting attitude.

In modern society, knowledge is supported by rather centralized systems. In postmodern society it will be supported by some IT mediated social networks. Knowledge is created, shared and evaluated on the social networks. The networks compete with each other. Network reputation becomes an important principle of competition. It is shown in Fig. 3.13.

3.6 Commitment and Learning Rationality

We formalized Axelrod's norm and meta norm games with weak coupled social learning dynamics. By using social learning dynamics we have analyzed the mutual and hierarchical commitment processes among agents. In the usual rational decision-making theory, we do not mention these types of social learning and commitment processes.

In other words we have introduced two types of extensions of the concept of rationality. The one is learning rationality, which means that social learning processes of agents are rational in a certain sense. In our social learning we do not assume perfect nor complete information. Agents do not interact such as in a person game. An agent decides by himself step by step depending on the average payoff information of each alternative at each step. Macro social learning consists of the step by step dynamics of the statistical decisions of agents.

The other is commitment rationality, which means that agents have mutual

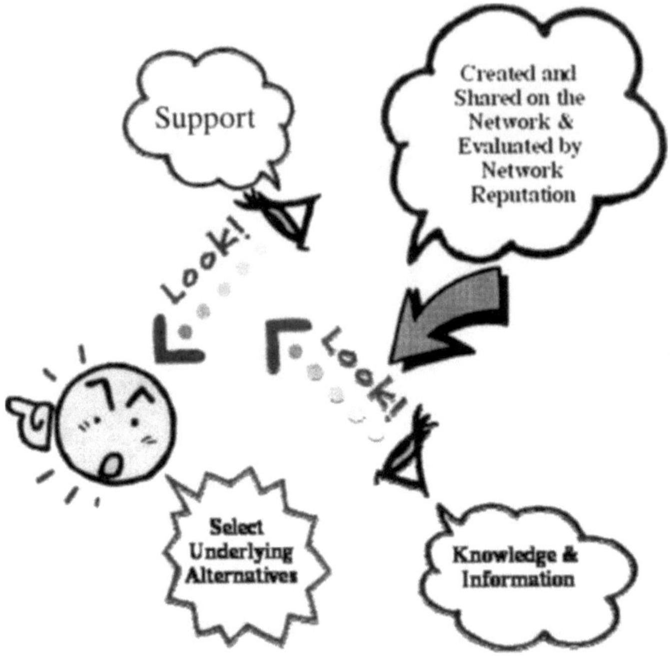

FIG. 3.13 Meta Commitment Through Knowledge and Information with Network Reputation

and hierarchical commitments such as "norm" or "meta norm". Why are commitments rational in a given situation? This question is a generalization of the one that asks why the norm collapses or does not collapse.

Robert Axelrod has introduced the concept of the meta norm [Axelrod, 1997, Ch.3]. But it brings up the next question of why the meta norm collapses or does not collapse. R. Gaylord and D'Andria mentioned this question in their book [Gaylord, 1998, Ch.5]. John Maynerd Smith also points out that the social norm is not always stable in his famous book, which is the origin of RD [Smith, 1982, Ch.13].

Our answer is simple. Mutual and self-commitments are rational if the alternative for the commitment is stable in the sense of social learning dynamics under the given social and political situation. The rationality of commitment comes from the stability of social learning dynamics. This is also an answer for the second-order social dilemma.

We also introduced the higher order centralized and decentralized commitment processes that affect an agent society as indirect regulation or control mechanisms. The theory of hierarchical commitments for an agent society is a new frontier for agent-based approaches. We need to develop the theory for designing our postmodern society.

For developing this area, both agent based heuristic simulation and theoretical analysis are important.

Chapter 4
The Tragedy of the Commons and Its Agent Based and Gaming Modeling

Aim of This Chapter

In this chapter we will focus on the tragedy of the commons and analyze the problem by using gaming simulation and agent based modeling. We will also investigate what kind of policies is essential to avoid this tragedy.

After the famous story of "the tragedy of the commons" by Garrett Hardin [Hardin, 1968; Schulz, 1994], many economists and sociologists have paid attention to the tragedy of the commons and its related topics. The story gives a strong metaphor for the problem of the use of common resources and in the case of the "tragedy of the commons", a common resource is exhausted by over use.

We compare the gaming simulation model of the "tragedy of the commons" with its agent-based simulation.

4.1 Gaming Simulation and ABM

Agent Based Modeling (ABM), a simulation method involving autonomous agents, has attracted attention in recent years, as a new and developing modeling method in the field of social science. Traditional methods that seek to understand, learn and analyze the complicated situations of socioeconomic systems through involving human beings as players in gaming simulation are also exploring new directions by incorporating computers and networks. In the past these two methods had very little to do with each other. However, it is very important to compare Agent Based Modeling with Gaming Simulation so as to determine ABM's validity and ability to describe actual socioeconomic systems. It is possible that a hybrid model combining both Gaming Simulation with human players and Agent Based Modeling will create a far wider range of possibilities.

Gaming Simulation is used across a broad range of fields including education, policy science, business studies and skills training [Greenblat, 1981,1988; Klabbers, 1996]. One of the roots of gaming is the analysis of political systems, including international relations, however, nowadays, the usage of gaming in this area has declined, at least in the U. S. and Europe. In policy think-tanks such as the Rand Corporation and the Brookings Institution there are now only a few gaming researchers and at the present time their focus is on ABM, a new policy simulation technique.

R. Axelrod emphasized the KISS principle, which is an acronym meaning Keep It Simple and Stupid, and using this concept as a modeling principle of agent based simulation maintained the robustness of the model [Axelrod, 1997].

On the other hand, throughout the world today gaming to solve problems and formulate policies for cities and communities is much-used technique. At a local level when designing cities, planning policy and solving problems in communities and regions, the gaming method maintains its position as a policy science and design science with a problem solving orientation.

In the field of business administration, business gaming is widely used as a means to help our understanding of business logic through a virtual game style and along with case studies they are an important means of providing practical training in business administration.

However, by combining ABM and gaming we would like to model real organizations and social phenomenon, as hybrid multi-agent systems consisting of both human and machine agents. This would enable the establishment of a system for analyzing, understanding and solving problems. Through this concept, a scenario would be created, that inevitably starts from the case example of a corporate organization and industrial structure, this would then in turn be analyzed and understood by creating a hybrid multi-agent system and could be applied to the creation of industrial policy and business analysis. It would also be possible to create such a social science research program.

In a similar way, gaming is used in the fields of sociology and organizational theory.

William Gamson's "SIMSOC" is the gaming simulation of an artificial society, so as to recreate the problems within a social structure [Gamson, 2000]. The hybrid multi-agent system will also open many new possibilities in this social field.

Next we will show an example of the relationship between GS and ABM.

4.2 Environment Management Gaming

Environment Management Gaming is a group gaming aimed at simulating a group of shepherds who graze sheep on a common pastureland.

Gaming simulation is not a computer simulation, it is a game style designed for human players. Gaming simulation is also different from a role playing game in which the roles of players are given by subjective description. In gaming simulation the role for each agent (player) is defined concretely and a player acts under the well-defined boundary conditions of the role. We adopt a certain gaming simulation and compare the results with (1) the theoretical and (2) computer simulation models.

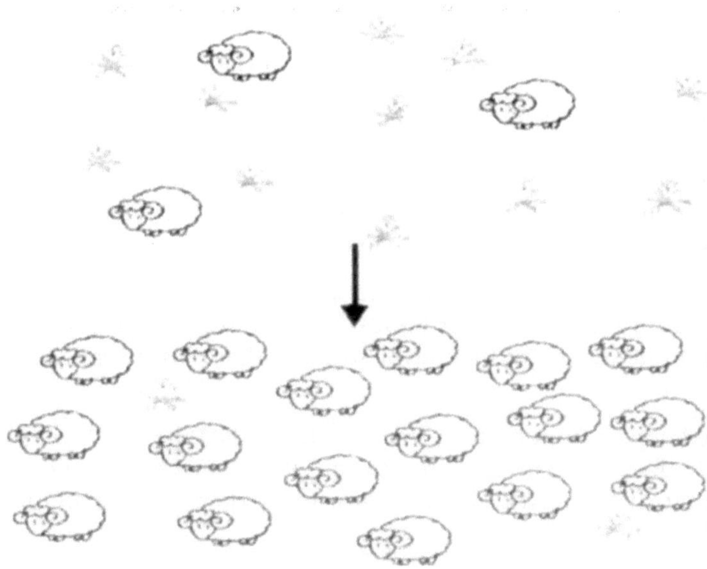

FIG. 4.2.1 Tragedy of the Commons

We assume that there are a limited number of shepherds, limited pastureland and being good shepherds they naturally want to multiply their flocks. However, because the land is limited a social and environmental dilemma is created among the agents, as well as a dilemma between individual and collective rationality.

A gaming facilitator coordinates all the transactions in the gaming. The players can perform a series of transactions with the facilitator: buy, sell, or rent sheep, and buy food or other goods. There is an abstract currency used for all transactions, the MOU (Money Unit). The goal of the players is to become wealthy by increasing the number of sheep they possess.

The gaming is divided into terms. At the end of each term the number of sheep owned by each player is multiplied by a reproduction rate. This rate depends on the total number of sheep in the pastureland. Players become wealthier if they increase their sheep, but the higher the total number of sheep in the pastureland the lower the reproduction rate. The players are posed a social and environmental dilemma.

4.3 How to Play Environment Management Gaming

A gaming facilitator plays Environment Management Gaming (EMG) with 5 to 150 or more players.

[Preparation]
The following items are recommended to use in the gaming.

(1) Counters that represent sheep.
(2) Counters that represent the unit of money (MOU).
(3) Counters that represent consumer goods (food, clothes, cars, houses, jewelry, etc.). An alternative to using counters is to write the number of goods on a card and place it in front of the player.

The players must consume 2 MOUs of food every gaming term. Cards and counters are used for other consumer goods. The number of the term is written on the card and put in front of the player so that other players can see the card at any time.

(4) A bookkeeping form to keep track of the rents. When a player rents sheep from the facilitator, he must return in double after two gaming terms (i.e., on the term after next). The facilitator must keep track of all the rents and obligations of the players.

[Initial Status]
At the beginning of the gaming, each player starts with three sheep and no cash.

[Players' Actions]
During the game, players sell sheep to purchase food and other goods. They can also rent or buy sheep from the facilitator by paying him money. Each sheep costs 2 MOUs. This price is constant throughout the gaming.

At the end of each term the facilitator calculates the reproduction rate and tells each player the resulting amount of sheep to start the next term. The reproduction rate varies from 1.0 to 2.0 but the players do not know the exact way this rate is calculated. At the end of each term the facilitator informs the player about the total number of sheep and the reproduction rate. When the reproduction rate is 1, the number of sheep remains the same, which means that the parent sheep do not die.

If the player rents some sheep, he must return in double at the beginning of the term after next. The player must pay for the sheep he buys in cash and the only way to raise cash is by selling sheep. The number of sheep to be sold or bought in a single term is limited to seven.

The players become bankrupt when they cannot return the rented sheep or cannot buy food. The players are allowed to sell rented sheep to pay for the food, but they cannot rent sheep before returning all the sheep previously rented (if any). After bankruptcy a player can return to the gaming with an initial three sheep.

Price of Goods
Sheep .... 2 MOU
Food .... 2 MOU
Clothes .... 2 MOU

Radio........................ 2 MOU
TV .... 4 MOU
Washing machine ... 4 MOU
Jewel .... 2 MOU
House ....20 MOU
Car...........................14 MOU

The player's record sheet is shown as follows.

Player's Record Sheet (basic version)

Player's name (   )   Date (/ /)   Term (   )

Resources at the Beginning of This Term
Number of sheep: (   ), Cash: (   ) MOU,   Goods: (   )MOU

Decision Making in This Term
Consumed food: (2)MOU (fixed)
Sheep sold: (   )   Sheep bought: (   )
Sheep returned: (   ) . . . . 2 × (number of sheep rented 2 terms back).
Goods bought: (   )MOU   Details:
Sheep rented (this term): (   )

End Term Calculation
Total number of sheep: (   )
= Number of sheep – sheep sold + sheep bought + sheep rented – sheep returned
Pastureland reproduction rate: (   ) . . . calculated by the facilitator and informed to all players for calculation
Sheep after reproduction: (   ),
Cash at the end of term: (   ) MOU
Total asset at the end of term: (   ) MOU
= Goods asset + cash + sheep (converted in MOUs)

4.4 The Facilitator's Guide of Environment Management Gaming

(1) Purpose of the Facilitator's Guide

The facilitator's guide gives the mechanism of the gaming to the facilitators. The players should not read this guide before playing the gaming.

This gaming, although played with basic rules, may raise unpredictable and complex situations. We can discover many valuable insights into economics, administration, and sociology through experiments based on this gaming.

(2) Assumption

In the gaming it is assumed that the increase in the total number of sheep in the pastureland degrades the environment, resulting in a decrease in the sheep's reproduction rate. The players do not know exactly how this rate changes.

(3) Preparation of Gaming

(1) Counters that represent sheep. Poker chips, color paper and coins can be used.
(2) Counters that represent the unit of money (MOU).
(3) Counters that represent consumer goods (food, clothes, cars, houses, jewelry, etc.). An alternative to using counters is to write the number of goods on a card and place it in front of the player.
    The players must consume 2 MOUs of food every term. Cards and counters are used for other consumer goods. The number of the term is written on the card and placed in front of the player so that the other players can see the card at any time.
(4) A bookkeeping form to keep track of the rents. When a player rents some sheep from the facilitator, he must return in double after two terms (i.e., on the term after next). The facilitator must keep track of all the rents and obligations of the players.
    The gaming can be played with a minimum of facilities: some counters, a white board and a calculator.

(4) Reproduction Rate

The reproduction rate cannot be shown to the players. At every term the facilitator must inform all the players about the total number of sheep and the corresponding reproduction rate. Given n players, the reproduction rate is calculated according to the number of sheep as follows;

Number of sheep	Reproduction rate
1 ~ 3n	2.0
3n + 1 ~ 3 × 2n	1.8
3 × 2n + 1 ~ 3 × 3n	1.6
3 × 3n + 1 ~ 3 × 4n	1.4
3 × 4n + 1 ~ 3 × 5n	1.2
3 × 5n + 1 ~ 3 × 6n	1.0

Example: Number of players = 5.

Number of sheep	Reproduction rate
1 ~ 15	2.0
16 ~ 30	1.8
31 ~ 45	1.6
46 ~ 60	1.4
61 ~ 75	1.2
76 ~ 90	1.0

Suppose a player has 3 sheep, with a total of 30 in the pastureland in this example. At the end of the term he will have $3 \times 1.8 = 5.4$. This number is round off to 5, and the player is given 2 sheep (2 counters) from the facilitator.

There are a lot of calculations to be made about rents, cash and goods. About 10 to 20% of the players will miscalculate during the gaming in usual. Using counters in the transactions facilitates the calculations and diminishes the errors.

The use of a bookkeeping form can also facilitate keeping track of the data. Even if the player has no knowledge about how to keep accounts, it is easy to explain how to read the data. The left hand of the form represents the actual assets and the right hand represents the details. The players should also be told that their reserves do not consist of the cash they possess.

The basic version of the player's record sheet may facilitate the conduct of the game, but is more error prone. The alternative use of the following bookkeeping form requires more time to fill out, which results in a term of approximately 30 minutes. A gaming with 10 terms would take 5 hours to complete. A total of approximately 6 hours is expected if we include the initial explanation and the debriefing. The institutional version of the gaming (with tax and subsidy) takes a total of approximately 12 hours to complete. The use of the simple form and the counters for sheep and cash speeds up the play and each term can be played in 10 to 15 minutes. A spreadsheet is more helpful for the calculation if it can be prepared.

Player's Record Sheet (bookkeeping version)

Player's name (   )   Date (/ /)   Term (   )

Balance Sheet at the Beginning of This Term: Stock

Debit Side		Credit Side	
Cash	( ) MOU	Equity	( ) MOU
Sheep	( ) MOU	Sheep rented in	( ) MOU
	( ) sheep	the previous term	( ) sheep
Goods	( ) MOU	Sheep rented 2 term back	( ) MOU
			( ) sheep
		Retained Earning	( ) MOU
Sum	( ) MOU		( ) MOU

Decision Making in This Term: Flow

Debit Side		Credit Side	
Cash	( ) MOU	Sheep sold	( ) MOU ( ) sheep
Sheep bought	( ) MOU ( ) sheep	Cash	( ) MOU
Food	(2) MOU (fixed)	Cash	(2) MOU Buy Food
Cost	(2) MOU (fixed)	Food Consumption	(2) MOU

Goods bought	( ) MOU		Cash	( ) MOU		
Sheep rented	( ) MOU	( ) sheep	Liabilities	( ) MOU	( ) sheep	
Liabilities	( ) MOU	( ) sheep	Sheep rented 2 terms back	( ) MOU	( ) sheep	
Cost: interest	( ) MOU	( ) sheep	Interest sheep	( ) MOU	( ) sheep	
Sum	( ) MOU			( ) MOU		

Remarks:
(1) The number of sheep returned in this term
= 2 × (the number of sheep rented 2 terms back)
(2) The number of interest sheep
= the number of sheep returned in this term
− the number of sheep rented 2 terms back

Calculation in This Term: Flow

Debit Side			Credit Side		
Cost: tax	( ) MOU	( ) sheep	Tax sheep	( ) MOU	( ) sheep
Subsidy sheep	( ) MOU	( ) sheep	Profit: subsidy	( ) MOU	( ) sheep
Sheep added	( ) MOU	( ) sheep	Added value	( ) MOU	( ) sheep

Remarks:
(1) Taxation and subsidy are used if we introduce an institutional design as is shown in the next section.
(2) Sheep added is determined as follows.

End Term Calculation
(1) Total number of sheep of a player: ( ) sheep
The total number of sheep of a player
= the number of sheep at the beginning of this term
− the number of sheep sold in this term
+ the number of sheep bought in this term
+ the number of sheep rented in this term
− the number of sheep returned in this term
− the number of tax sheep in this term
+ the number of subsidy sheep in this term

(2) Pastureland reproduction rate: ( )
Pastureland reproduction rate is calculated from the sum total number of sheep in the pastureland at the end of term by the facilitator and informed to all players for their calculation

(3) Sheep after reproduction: ( ) sheep
The number of sheep after reproduction is calculated from the total number of sheep of each player and the reproduction rate. The number of sheep of the player is multiplied by the reproduction rate and rounded off to an integer number

(4) Sheep added: ( ) sheep
Sheep added
= The number of sheep of the player after reproduction
– The total number of sheep of the player before reproduction
= Value added

(5) Cash at the end of term: ( ) MOU
Cash at the end of term
= Cash at the beginning of term
+ Sheep sold (converted in MOUs)
– Sheep bought (converted in MOUs)
– Food Consumption (2 MOU fixed)
– Goods bought

(6) Total asset at the end of term: ( ) MOU
Total asset at the end of term
= Goods at the end of this term
+ Cash at the end of this term
+ Sheep at the end of this term (converted in MOUs)

(7) Retained earning: ( ) MOU
Retained earning at the end of term
= Retained earning at the beginning of term
+ Value added (converted in MOUs)
+ Subsidy sheep (converted in MOUs)
– Interest sheep (converted in MOUs)
– Tax sheep (converted in MOUs)
– Food consumption (2 MOU fixed)

(5) Gaming Termination

The players should not be informed about when the gaming is to terminate, so as to avoid the players making terminal strategies during the gaming. Normally it takes 7 to 8 terms for the gaming to stabilize (oligopolistic market with several players going bankrupt). Therefore a total of approximately 12 terms is recommended for the gaming. The institutional version (with tax and subsidy) of the gaming also requires about 12 terms because it takes some time for the players to learn about the rules and the complex combinations of strategies that may arise.

(6) Institutional Design

Tax: players must pay tax if they have more than 7 sheep. The tax is paid in sheep. The number of sheep owned by the player is multiplied by the tax rate, the result is then rounded off to an integer number, and the corresponding number of sheep must be paid to the facilitator at the end of each term.

(1) Simple taxation: tax sheep = (Number of sheep – 6) × tax rate. (Decimal numbers are rounded off to integers.)
(2) Progressive taxation: The tax rate varies according to the number of sheep.

Example:

Number of sheep	Tax rate (%)
0 ~ 6	0
7 ~ 10	10
10 ~ 15	20
16 ~	30

Subsidy: The collected tax is distributed to the players who have less than five sheep. If the number of taxed sheep cannot be divided exactly among the players, the remaining sheep are pooled for the subsequent terms.

One can play different gaming with combined rules: with taxation only, with taxation and subsidy, with simple taxation of 10%, with simple taxation of 20%, and with progressive taxation. Another interesting variation for the gaming is to introduce polls to decide on increasing the tax rate or changing the taxation rules.

(7) Debriefing

Discuss about the impoverishment of the society when there is a social dilemma and no institutional measures. Discuss about the process where monopolistic situations arise. Confirm that subtle differences in strategies can lead to big differences in the results, and then discuss the results relating them to the initial strategies.

Debate about the ways to improve society and then play a variant of the gaming, introducing the debated institutional rules.

In the institutional version we can often observe strategies that try to lessen the tax burden and the tax being low confronted with strategies that are in conform with tax paying to be wealthier. Discuss how this confrontation changes as a consequence of taxation rules and subsidies.

4.5 Human Gaming Simulation of the Commons

We compare the following results of human gaming simulation with the results of agent-based simulation shown in the next section under the same condition. The same types of activities are found in both cases. Cases of gaming simulation by human players are shown as follows. Figure 4.5.1 indicates the change of total assets which are measured by MOU (Money Unit). Player No. 6 and No. 8

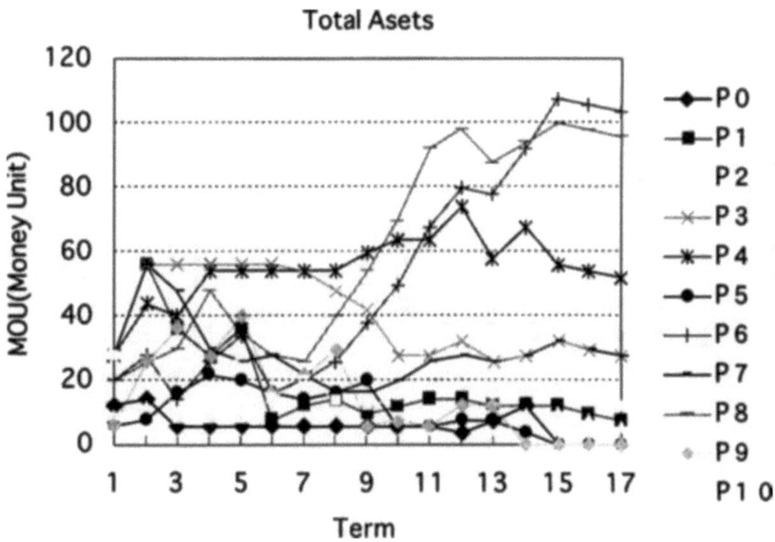

FIG. 4.5.1 Total Assets of the Players

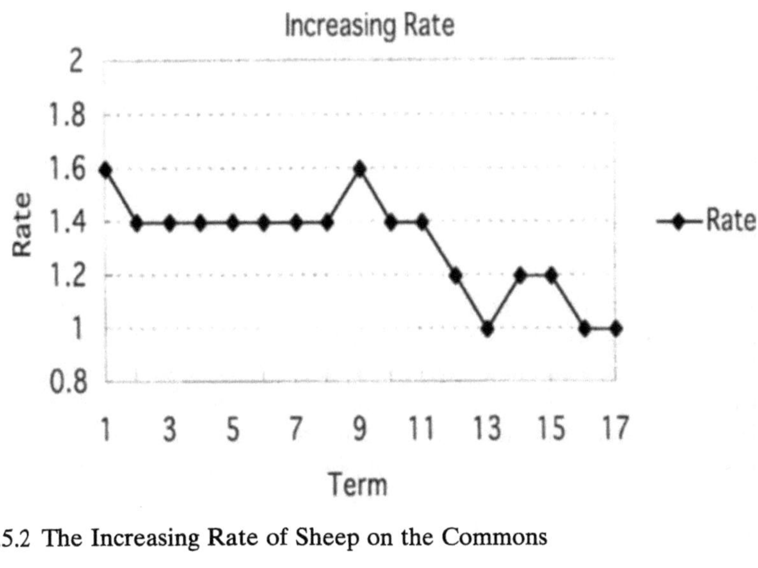

FIG. 4.5.2 The Increasing Rate of Sheep on the Commons

became rich and won in this gaming, while the other players became poor and the gap between rich and poor became wider. On the other hand Fig. 4.5.2 indicates the change of the production rate of the commons and as a result only a tiny handful of shepherds survive and the commons become non sustainable.

These cases show the commons with a policy of no indirect control. A number of dominant agents survive in the commons and the commons become unsustainable with a few dominant players becoming the winner in the long run.

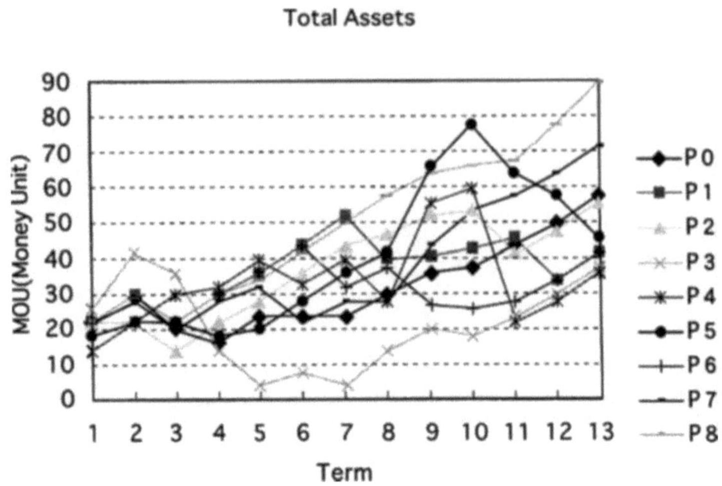

FIG. 4.5.3 The Total Assets of Players with Tax and Subsidy

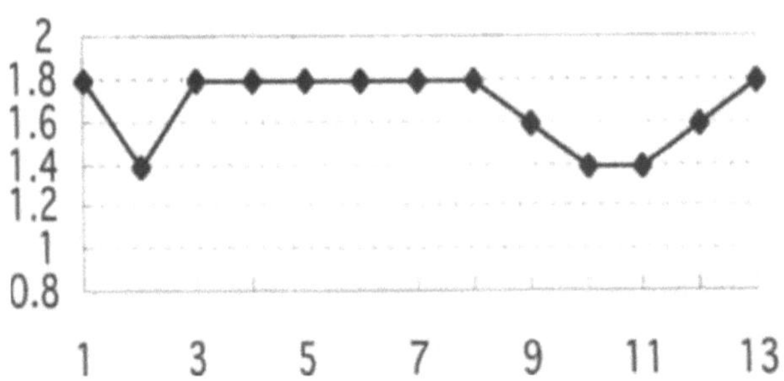

FIG. 4.5.4 Increasing Rate of Sheep with Tax and Subsidy

Institutional cases show the commons with a wealth tax on rich agents and a subsidy for bankrupts. In this case the commons become sustainable and a variety of activities are observed. There is a strong repeatability in this gaming.

We learn about the meaning of an institutional framework that is regulating a social complex system through the experience of gaming simulation. Figure 4.5.3 shows the change of the total assets of players with a tax and subsidy policy. Such indirect control gives rise to a variety of competitive behaviors among players and Fig. 4.5.4 shows that the commons became sustainable.

4.6 Agent Based Simulation and Classifier

Through gaming simulation we characterized the tragedy of the commons, however, this analysis by gaming is limited because of the ability of the human agents. We can now introduce machine agents to play the game and analyze the variety of gaming by agent-based simulation on a computational reality.

TABLE 4.6.1 Bit Coding of Condition Part and Action Part

Number of Sheep	Condition Part	0	1–5	6–10	11–15	16–25	26–35	36–50	51–
Cash (MOU)	Condition Part	0	1–3	4–8	9–15	16–22	23–30	31–40	41–
Number of Sheep Borrowed	Condition Part	0	1	2	3	4	5	6	7
Number of Goods	Condition Part								
Number of Sheep to Borrow	Action Part	0	1	2	3	4	5	6	7
Total Number of Sheep in Commons	Action Part	0–15	16–30	31–45	46–60	61–75	76–90	91–100	100–
Number of Sheep to Sell	Action Part	0	1	2	3	4	5	6	7
Coding	Bit Sequence	000	001	010	011	100	101	110	111
Shopping (Goods to Buy)	Action Part	0	1						
Coding	Bit Sequence	0	1						

TABLE 4.6.2 Example of Rules

(1) Condition Part

Number of Sheep	Cash (MOU)	Number of Goods	Number of Sheep Borrowed	Total Number of Sheep in Commons
010	##0	0#	00#	0#0

(2) Action Part and Weight

Number of Sheep to Sell	Number of Sheep to Borrow	Number of Goods to Buy	Weight of Rule
001	011	1	100

For this purpose we shall introduce the classifier system to describe the action rules of a shepherd. The classifier system was originally introduced by J. Holland and is a form of production system with reinforcement learning. For describing the production rules in a classifier system we have to define the condition part and the action part by using bit coding. The condition part is used for decision making as information by a player and the action part shows the selection of alternatives open to a player. In this model we use the number of sheep, cash, borrowed sheep, total number of sheep in the common for condition parts and the number of sheep to sell, number of sheep to borrow and number of goods to buy for the action part. We encode a rule by a fourteen binary digit for the condition part and by a seven binary digit for the action part.

The coding rules of the condition and action parts are given as follows as a sequence of bits. Table 4.6.2 shows an example of a rule. Initially we assign 0, 1 or # randomly to each digit of the condition part of a rule. We also assign 0 or 1 randomly to each digit of the action part of a rule. All rules have the same initial value 100 as a weight of rules. Each player has 50,000 rules.

4.7 Agent Based Simulation of the Commons

In the gaming simulation we characterized the tragedy of the commons in the game. However the analysis by gaming is limited because of the ability of human agents. Instead we can introduce machine agents to play the game. We can analyze the variety of the gaming by agent based simulation on the computational reality.

For this purpose we introduce the classifier system to describe the action rules of a shepherd. In the following case five shepherds are living on the pasture

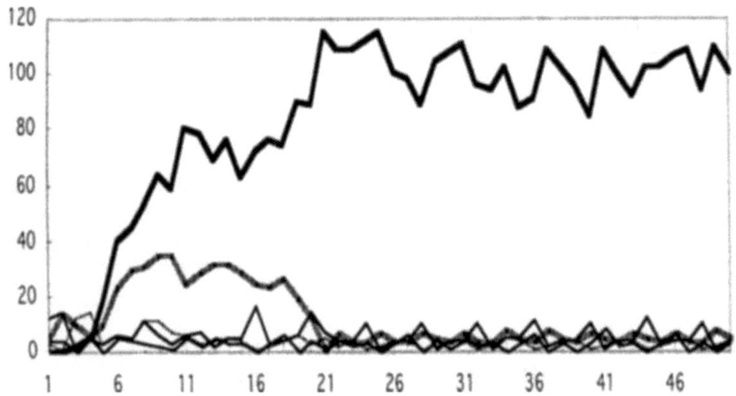

FIG. 4.7.1 The Number of Players' Sheep: Agent Based Simulation

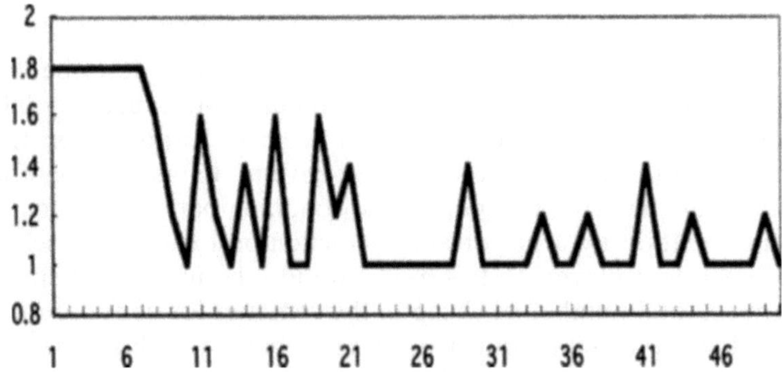

FIG. 4.7.2 Increasing Rate of Sheep on the Common: Agent Based Simulation

land. In a common with an investment mechanism, the common is not sustainable and a sheep baron and poor peasants appear, giving rise to the extremes of enduring poverty and dynastic wealth within the model.

We compare the previous results of gaming simulation with the results of agent based simulation under the same condition. The same types of activities found with human based gaming simulation are seen and the results of agent based simulation by machine players are shown in Figs. 4.7.1 and 4.7.2.

In this case one or two dominant agents survive and the commons also become unsustainable. We notice that there are two macro functional satisfaction conditions in this system. In other words there are 2 axis of indirect control by changing the boundary conditions.

(1) Sustainability of the commons
(2) The gap between rich and poor

By what means is it possible to satisfy these two macro functional conditions at the same time? Can one or more structural parameters for generic bifurcation of this agent society be found in order to satisfy these two macro requirements?

Our policy goals are sustainability of the common and a reduction in the difference between poverty and wealth. Two types of indirect control method are introduced.

(1) The one is a wealth tax on rich agents.
(2) The other is a subsidy for bankrupts.

The results are shown in Figs. 4.7.3 and 4.7.4. In this case of a wealth tax on the rich, the commons became sustainable and one or several dominant agents survive on the commons.

With a wealth tax on rich agents and a subsidy for bankrupts the commons become more sustainable and a variety of activities are observed.

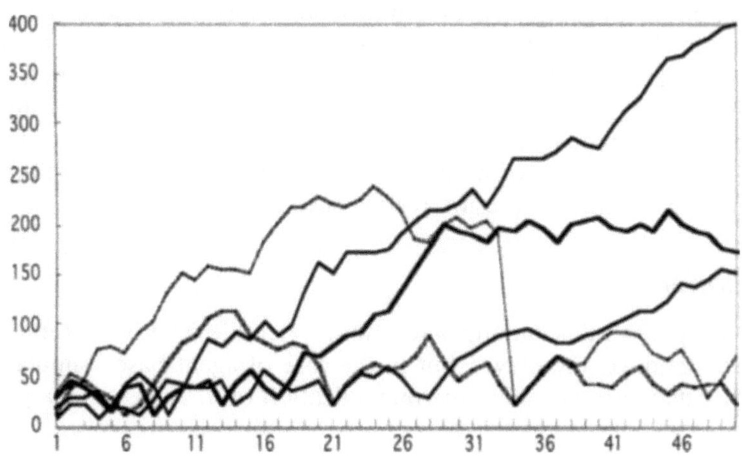

FIG. 4.7.3 The Results of a Wealth Tax

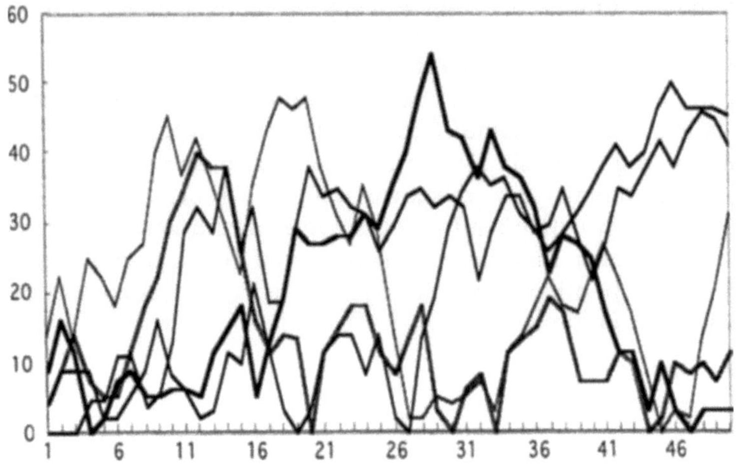

FIG. 4.7.4 The Result of a Wealth Tax and Subsidy to the Poor

4.8 Agent Based Simulation of the Commons

4.8.1 Agent Based Simulation and Fact Finding

In Chapter 1 Fig. 1.2 we have already shown the methodological triangle by which we insist that experiment by agent based simulation is important because of its fact-finding ability.

In the previous simulation we found that in the long run and in almost every case a strong disparity in wealth occurred. This gaming was originally designed for teaching an environmental problem in a sociology class and in the case of human gaming simulation it is difficult to repeat the game a large number of times and therefore we did not realize that there is such a serious gap in the first four or five steps of gaming. After the agent based simulation we changed the standard number of repeated gaming to a minimum of ten times. Moreover we can easily find this disparity of wealth in the historical commons. In the following simulation we found a more interesting scenario and result.

4.8.2 Change of Value

A player's purpose in this gaming is to attain a rich life. The interpretation of a "rich life" depends on the player, but in the case of agent-based simulation we give an exact definition meaning the maximization of total assets of a player as follows.

$$\text{Total Assets} = \text{Total Cash} + (\text{Number of Sheep}) \times 2 + (\text{Number of Goods}) \times 2 \qquad (4.8.1)$$

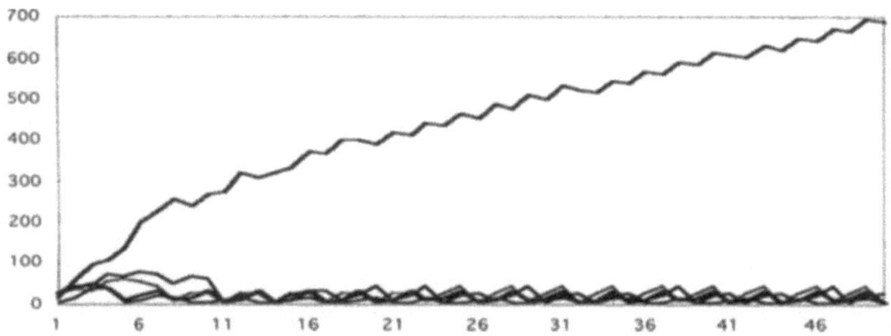

FIG. 4.8.1 Total Assets with Modified Values

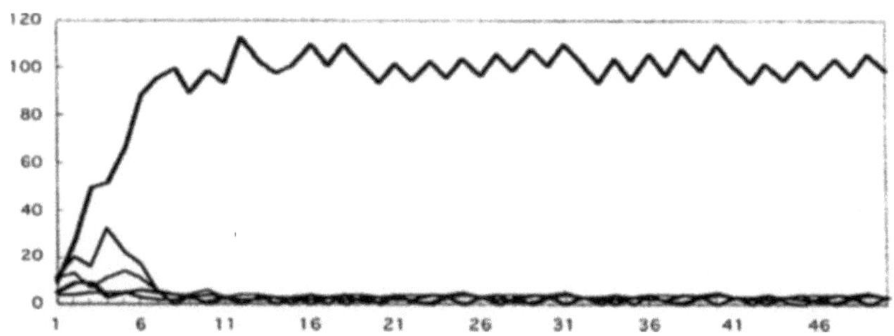

FIG. 4.8.2 The Number of Sheep with Modified Values

In this simulation we assume that we can buy food and goods as a consumer. Foods have been consumed. Goods are accumulated and then counted as total assets.

In the case of human gaming we originally introduced a greater variety of goods such as TV, jewelry, cars and houses all with different prices. In the present version we assume only one type of general item with a price of 2 MOU. Then (4.8.1) denotes the total value of assets of a player.

It is important to determine whether the result of gaming will change or not when agents have different aims to attain. It is difficult to design a gaming experiment in which players change the value system as an experimental control parameter, however, it is easy to change the value system in agent based simulation. We can change the evaluation function of action rules for this purpose and at first we change the value system for the society where the value of goods is high. For this purpose we count the value of goods as (Number of Goods) × 10 for example.

Then the result of simulation is shown as follows. A sheep baron does arise in this case as is shown in Fig. 4.8.1, where the X axis denotes the time steps of the gaming and the Y axis denotes the total assets measured by MOU (money unit). Figure 4.8.2 shows the change of the number of sheep of each player. Figure 4.8.3 shows the change of the production rate of the commons and there is no sustainability in this commons.

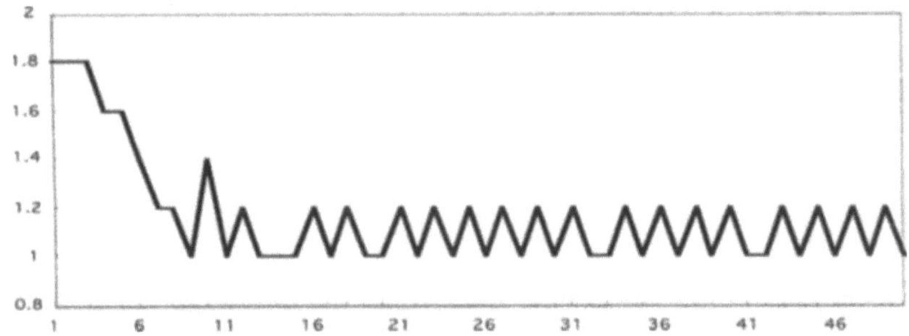

FIG. 4.8.3 The Reproduction Rate of the Commons with Modified Values

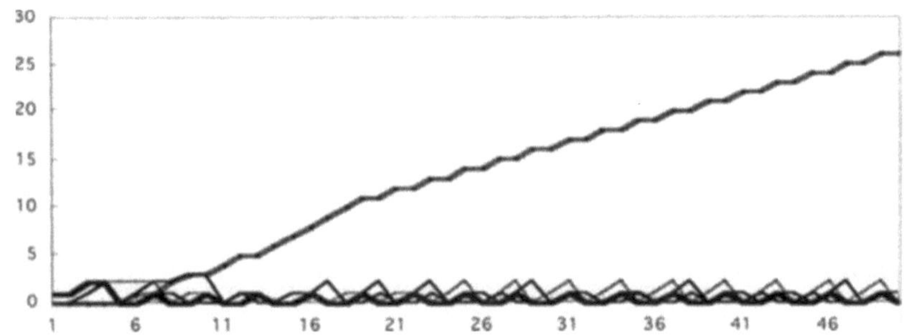

FIG. 4.8.4 The Number of Goods with Modified Values

At first we intended to control the sustainability of the commons and the gap between the rich and the poor by changing the value system in this simulation. This approach failed, however, we noticed that the strategy of the winning player was unique, with the winner at first trying to become a champion under this modified value system. In this stage he only tried to increase his sheep, but after becoming champion of the commons in the next stage he then tried to increase his goods as is shown in Fig. 4.8.4. We call this strategy the "Snob Strategy" where he first tried to become rich and then raise his cultural or other values by using his wealth. The "Snob Strategy" is a very popular strategy in real life and it can be reproduced by this simple learning mechanism. Of course our system is simple and there is no hierarchical commitment or other learning structure.

4.8.3 Tax and Subsidy as Institutional Mechanisms of Agent Society

We introduced tax and subsidy in this gaming as a classical institutional framework to control sustainability of the environment and the social disparity in wealth. At first we introduced tax in proportion to the number of sheep and the results are shown in Figs. 4.8.5 and 4.8.6. The commons became sustainable but there is

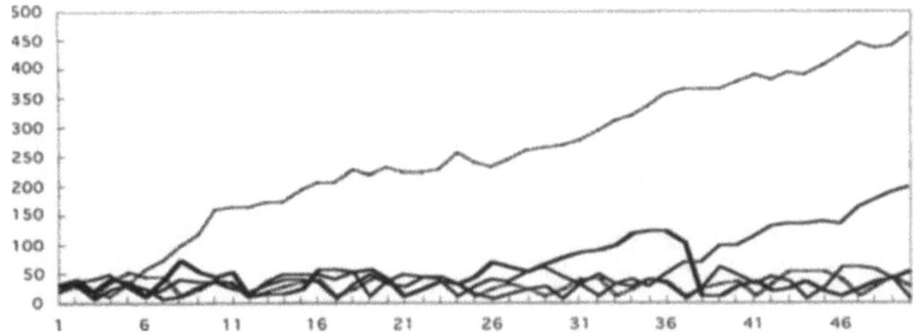

FIG. 4.8.5 Total Assets with a Wealth Tax

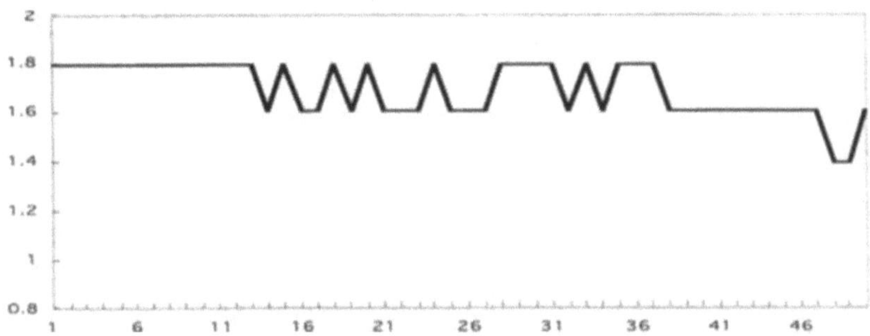

FIG. 4.8.6 Increasing Rate with a Wealth Tax

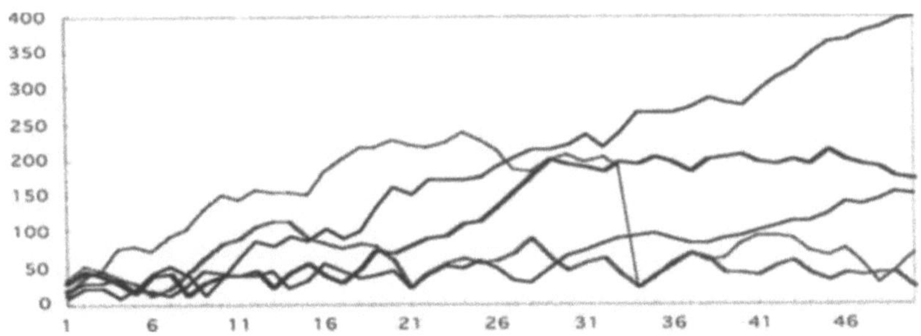

FIG. 4.8.7 Total Assets with Tax and Subsidy

a gap between the rich and the poor. There is a winner and a second winner in this commons.

Tax controls the increasing rate of sheep, because the winner has to remove sheep from the commons as a tax payment. This creates the opportunity for a player to become the second richest, but many players became bankrupt.

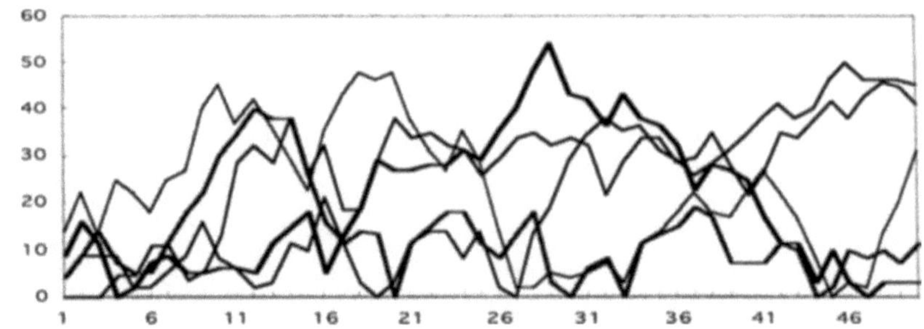

FIG. 4.8.8 Total Number of Sheep with Tax and Subsidy

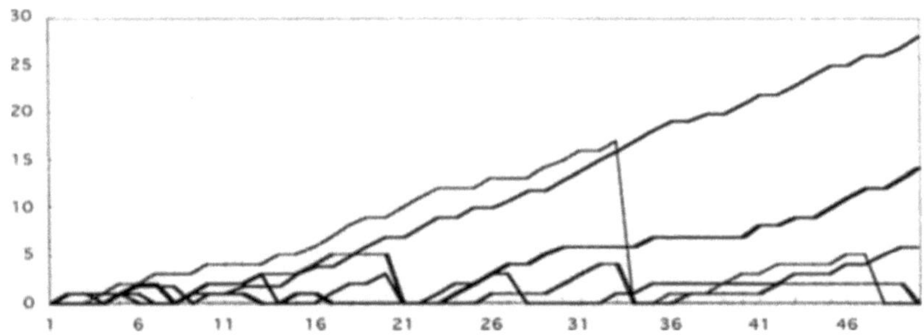

FIG. 4.8.9 The Number of Goods with Tax and Subsidy

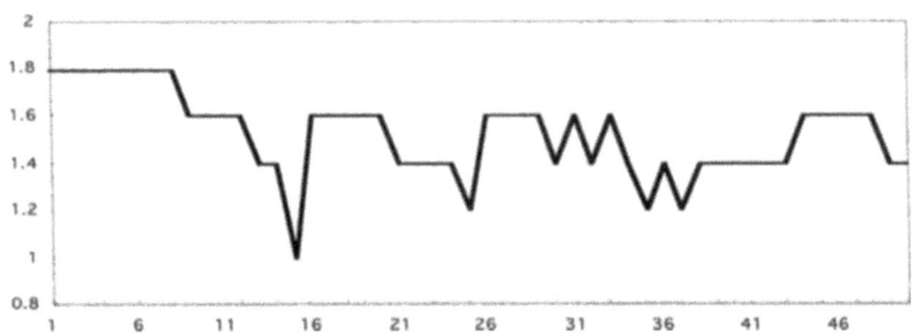

FIG. 4.8.10 The Increasing Rate with Tax and Subsidy

Next we introduced a subsidy mechanism to redistribute wealth.

A player who does not have more than 30 MOU as assets at the end of each term received sheep as a subsidy from the total tax on a per capita basis. But the upper limits of the subsidy to each agent is five sheep. The results of simulation are shown in Figs. 4.8.7, 4.8.8, 4.8.9 and 4.8.10, with a variety of competition in this commons and sustainability is maintained.

We have shown that tax and subsidy are control parameters in this system and the structure of the commons change under the influence of these parameters.

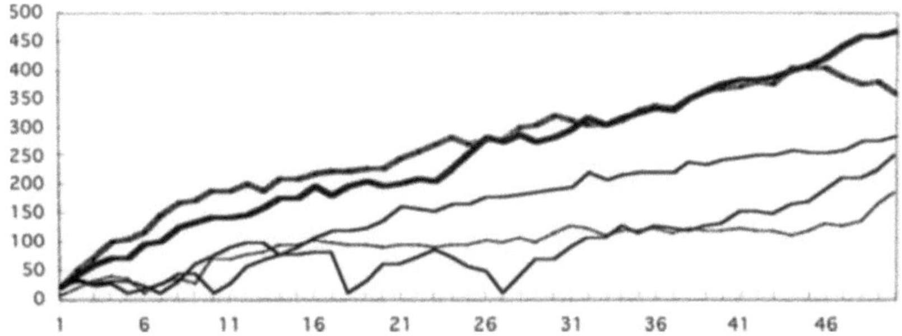

FIG. 4.8.11 Total Assets with Tax, Subsidy and Education

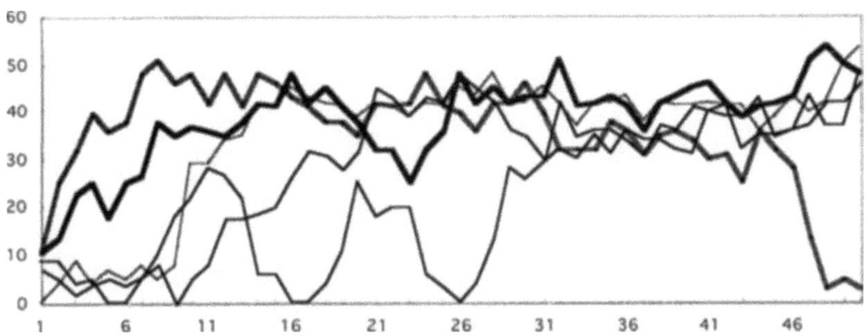

FIG. 4.8.12 The Total Number of Sheep with Tax, Subsidy and Education

In other words there is an indirect control mechanism in this commons. Of course we have not formulated the commons as a dynamical system in mathematics, but the concept of a steady state of a system, structural change, social learning and indirect control, as introduced in the previous chapter, can be applied to this system.

The learning introduced in this section is a typical multi-agent reinforcement learning, by which an agent learns effective rules to attain his aim.

The simulation always gives the same types of results where the winner might change but the structure is the same. This means that the result is robust and also coincides with the results of human gaming. But there is a slight difference between agent-based simulation and human gaming, in the machine learning case agents learn strategy by reinforcement learning from reward. They do not learn from each other, as a result there are many cases of bankruptcy in agent-based simulation, even in the tax and subsidy case.

In the case of human gaming players can learn successful strategies from each other and then as shown in the previous section there are few cases of bankruptcy in the tax and subsidy case.

Finally we introduced the effect of education where we assume that agents can learn effective rules from successful agents [Terano, 1998; Takadama, 1999] and the results are shown in Figs. 4.8.11, 4.8.12, 4.8.13 and 4.8.14.

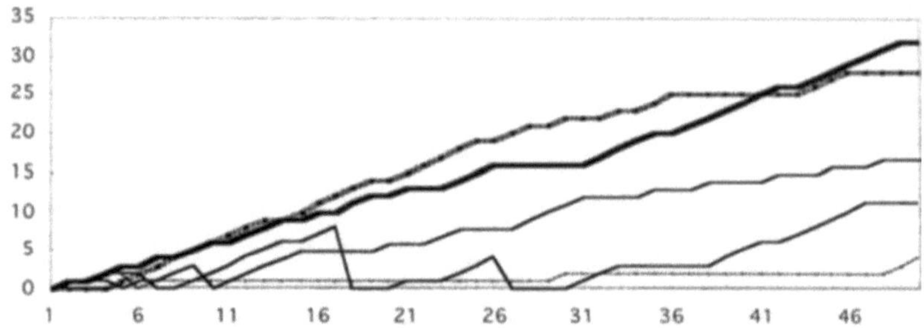

FIG. 4.8.13 The Number of Goods with Tax, Subsidy and Education

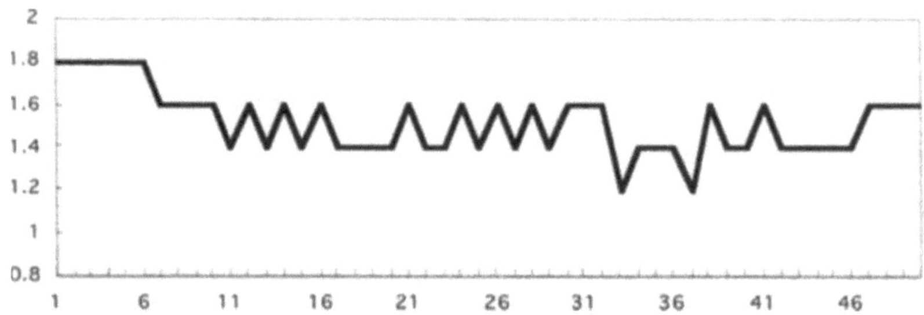

FIG. 4.8.14 The Reproduction rate with Tax, Subsidy and Education

The number of bankruptcy cases has decreased in this simulation and society became more stable and fair. As a result in this simulation we have reconfirmed well-known principles from real life such as monopolies not being permitted, the chance of restoration is important and education and learning from others are important to avoid failure caused by a lack of knowledge.

4.8.4 Random Strategy and Docking Method

We have compared the result of our agent-based model with the result of human gaming. In agent-based simulation docking method, comparing with another simulation method for checking the validity of original simulation, is recognized as very important [Axelrod, 1997]. Thus we introduce another agent-based simulation model by using Stella, a system dynamics simulator, for checking the validity of our agent-based simulation. The model is simplified and there is no learning of strategies. We use random strategy with certain constraints in this model. Even in this simple model there happen lock-in and unsustainable situation in the pastureland.

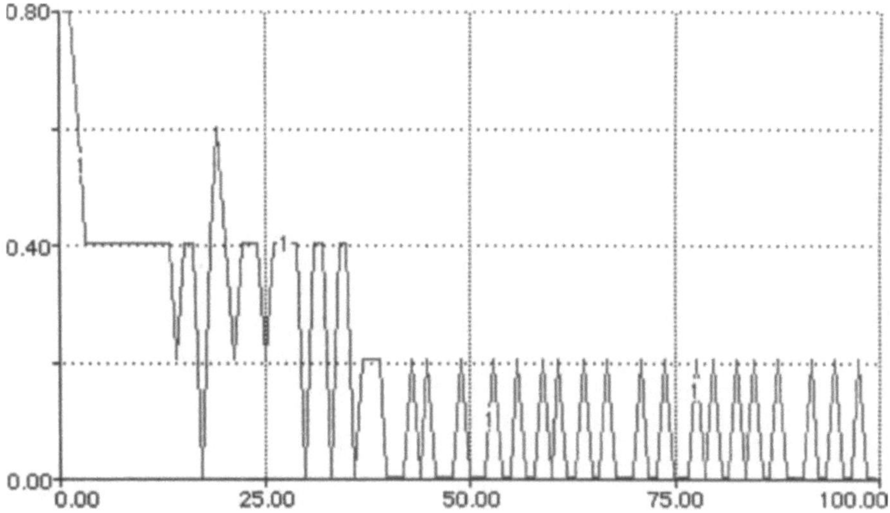

FIG. 4.8.15 Increasing Rate by Random Strategy Model

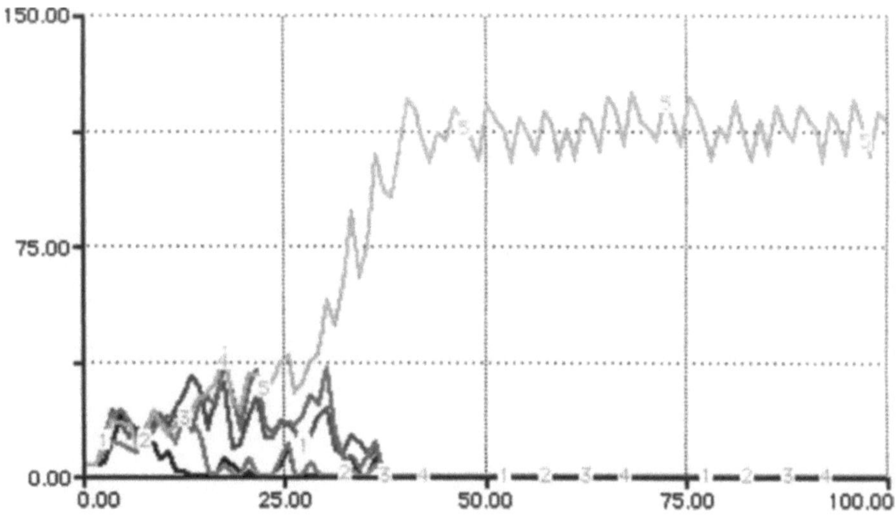

FIG. 4.8.16 Number of Sheep by Random Strategy Model

The results are shown in Figs. 4.8.15 and 4.8.16. We can not forecast who become a winner in the pastureland. If we change a random seed then the winner change drastically. Thus the selection of a winner depends not on the strategy but on the selection of a random seed. But the destiny of the pastureland is miserable and the result is robust.

Chapter 5
State Space Complexity and Exchange Algebra

Purpose of This Chapter

In this chapter, the mathematics of a state space description for economic exchange are considered. This mathematics becomes the basis for a micro and macro state description of an economic system. This axiom gives the basis to analyze the state complexity of an economic system and as a result, some characteristics of economic exchange become clear.

First, a bookkeeping system is characterized by the following two axiomatic systems. The one is called redundant algebra and the other is called exchange algebra.

For example, the credit side and the debit side of a bookkeeping system are characterized as equivalence classes in exchange algebra. In addition, it is shown that there are only two equivalence classes.

It was thought to be self-evident in a bookkeeping system that both sides balance. This is also shown to be a theorem from our axioms.

Moreover much useful knowledge can be obtained from this bottom-up approach by using an algebraic point of view. For example field descriptions of economic systems and multi dimensional bookkeeping systems can be formulated from our framework.

5.1 Introduction

In this chapter we analyze the structure of an internal model for a state description of an economic agent from an algebraic point of view.

Economics has established a rational approach to the decision making of an agent. The agent-based approach has introduced the learning and evolution mechanism of agents. It has become clear that analysis of an internal model for a state description of an agent is unavoidable in order to analyze the complexity of an economic agent.

An internal model of a state description becomes the basis for constituting other internal models such as strategy, evaluation or causal relations. If an agent does not have an appropriate internal model of the state description of an agent, then it is difficult to evaluate the actions of the agent appropriately and also difficult to understand causal relations to its environment.

The state description of economic systems by stock variables is as important as the flow description of the system. An economic system as a dynamical system has a remarkable characteristic in the model constitution. All the important macro variables for economics such as GDP, investment, consumption and savings are flow variables. In other words an economic system is a system in which flow variables play unique roles in creating its characteristics.

Actually, in the history of economic models, a state variable for stock description has not been used properly. One of the reasons is that economic theories put too much importance on "an equilibrium" concept. Keynesian economics pays more attention to a short-period equilibrium. It is rare to use stock variables even if they are used for acceleration principles. However stock variables are always used as a state variable in physical systems.

In a recent study of the reconstruction of macroeconomics from a microeconomic foundation a concept of state variable is used. The state concept does not analyze the relation between the micro state description of each economic agent and the macro state description of a national economy. We give an algebraic formulation of the state concept that describes the stock of an economic system and its change.

For a long time the concept of economic exchange has been considered. However the focus is on how the equilibrium of supply and demand is achieved by exchange. This is a trial to explain what kind of exchange occurs with the utility function. The focus is on what kind of exchange decision making is carried out.

Otherwise there exists a system of bookkeeping, which describes the traded properties of economic exchange. This system offers an inter subjective state space for all economic agents. The bookkeeping system gives an inter subjective internal model to an economic agent for describing its economic activities.

A description of economic activity by bookkeeping is the core of any accounting system. Economic activities as transactions are described by a bookkeeping system.

An accounting description constructs an intersubjective internal model of the economic state of a firm. A concept of accounting measurement shows that an accounting system is a cognitive framework. If the construction of this internal model is different, then the strategy of an agent's activity will change. The strategy of a firm depends on accounting standards as an inter subjective internal model.

In addition to business transactions, various economic activities such as production and consumption can be described by a double-entry bookkeeping system. Double-entry bookkeeping developed as a system that describes economic activities by exchange. However, there is a limit to its description.

For example a materials account is indispensable for analyzing the economic activity of manufacturing firms, but a materials account cannot be described by the usual double-entry bookkeeping system.

An extension of the bookkeeping system for a materials account was tried as a multi dimensional bookkeeping system [Ijiri, 1990], but no strong mathematical foundation was established.

State construction of a national economy derived from the composition of a bookkeeping description of the economic agents has not been developed, except for a trial approach with a matrix bookkeeping system by R. Mattessich [Mattessich, 1995a; Koshimura, 1976].

We extract the algebraic deep structure, which exists behind the repeatable surface structure of a double-entry bookkeeping system for this extension [Deguchi, 1984B, 1986].

This abstraction is a so-called structuralist approach. By using an algebraic formulation of the transaction, it is possible to formulate an exchange of multi dimensional bookkeeping system for a materials account and to give an algebraic formulation of the system of national accounting (SNA). In this chapter we give the algebraic extraction of bookkeeping and show the basic characteristics of the system with a step-by-step abstraction.

In the first step of the abstraction, an algebraic system called accounting algebra is extracted. This algebra is constructively introduced from an accounting system. This algebra requires an accounting system to satisfy the axioms of exchange.

In this algebra the debit and credit sides of the bookkeeping system are characterized as an equivalence class, containing only two classes. This accounting algebra is characterized as an algebraic system generated from a base set.

In the second step the algebraic system is completely characterized in an axiomatic way. This axiom is then divided into two parts.

At first algebra that does not use negative numbers in its description is characterized by an axiom. This algebra is called redundant algebra. Next we add an axiom of exchange that characterizes the exchange of a double-entry bookkeeping system.

Then an algebraic system called exchange algebra is introduced [Deguchi, 1990, 2000b]. The well-known characteristics of a bookkeeping system can be deduced form the exchange algebra.

A multi dimensional and multi subjects bookkeeping system is deduced by the natural extension of exchange algebra. Of course there was an attempt at an axiomatic description of the concept of exchange.

R. Mattessich pointed out several axiomatic approaches, which include our approach in his book [Mattessich, 1995b]. However other approaches except ours try to characterize not the state structure of bookkeeping but the accounting standard itself as a decision-making principle. Only our approach gives an algebraic extraction of a bookkeeping system and its axiomatic characterization.

5.2 Axiomatic Foundation of Economic Exchange

First we introduce an algebraic system to describe economic exchange and the basic characteristics are examined. The algebraic system that we shall introduce is an abstraction of the double-entry bookkeeping system. At first a constructive definition of accounting vector space is given as an algebraic system. Next we show that the algebra becomes an abstraction of a double-entry bookkeeping system. Furthermore this accounting vector space is abstracted and it is characterized by abstract axioms. First title of account gives a set of base as follows.

[Definition 5.2.1] A Set of Bases

Let $\Lambda = \{e_1, e_2, \ldots, e_n\}$ be a set of suitable title of account.

$[\Lambda]$ denotes the commutative free semi group with a unit element 0, generated by a summation of elements of Λ.

Then Λ is called a set of base and e_i is called a base of account.

What we expressed as bases is an abstraction of an account such as <cash> or <equity>.

[Example 5.2.1]

Let $x = 3e_1 + 2e_2 + 5e_3$. x is generated by $\Lambda = \{e_1, e_2, e_3\}$. Then x becomes a element of $[\Lambda]$.

For this algebra mathematical operations such as $3e_1 + e_1 = 4e_1$ and $4(3e_1 + 2e_2) = 12e_1 + 8e_2$ are defined.

In other words, $[\Lambda]$ is an algebra with external multiplication "$*$" : $N \times [\Lambda] \rightarrow [\Lambda]$.

For example $a * (be_1 + ce_2 + de_3) = abe_1 + ace_2 + ade_3$.

[Example 5.2.2]

Let $x = 500e_1 - 300e_2$. This means an income of cash 500 yen and expenses of apple 300 yen by a single-entry bookkeeping system, where e_1 = <cash> and e_2 = <apple>.

In the natural sciences a negative number is usually used for describing the decrease of stock. However a negative number is not used in a double-entry bookkeeping system for describing the decrease of stock, instead a double-entry description is used.

This is an essential point for a state description of economic exchange. To clarify this feature we introduce the concept of dual base and extension of base from an algebraic point of view.

[Definition 5.2.2] Dual Base and Expansion of Base

Let $\Lambda = \{e_1, e_2, \ldots, e_n\}$ be a set of bases, which consist of title of accounts.

Then new base $\hat{\ }e_i$ is introduced from $e_i \in \Lambda$, which is called a dual base of e_i.

$\hat{}\Lambda = \{\hat{}e_i \mid e_i \in \Lambda\}$ is called a set of dual bases.
$\Gamma = \Lambda \cup \hat{}\Lambda$ is called a set of extended bases of Λ.

[Definition 5.2.3]
F: $\Lambda \to \hat{}\Lambda$ is defined by $f(e_i) = \hat{}e_i$, which is a function to determine natural correspondence from Λ to $\hat{}\Lambda$

It is clear that $f^{-1}(\hat{}e_i) = e_i$ is one to one and onto function from Definition 5.2.2.

Thus we introduce "^" as a function from Γ to Γ instead of f^{-1} and f, where $\hat{}\hat{}e = e$ is holds.

In the same way $\hat{}x$ is introduced as a function from $[\Gamma]$ to $[\Gamma]$.

For example $\hat{}(3e_1 + 5\hat{}e_1 + e_2) = 3\hat{}e_1 + 5e_1 + \hat{}e_2$ holds.

[Definition 5.2.4] Set of Accounting Bases
Let $\Lambda = \{e_1, e_2, \ldots, e_n\}$ be a set of bases.
Let Ω = {<assets>, <liability>, <equity>, <revenue>, <expense>}.

If there exist onto function k: $\Lambda \to \Omega$, which classify elements of Λ, then Λ is called a set of accounting bases.

The definition shows that an account list can be classified by the elements of Ω.

[Definition 5.2.5] Accounting Vector Space and Accounting Algebra
Let Λ be a set of accounting bases.

Let $\Gamma = \Lambda \cup \hat{}\Lambda$ be a set of extended bases of Λ.

Then $[\Gamma]$ is called an accounting vector space, which consists of accounting bases Λ. An element of $[\Gamma]$ is called an accounting vector.

An accounting vector space is not a linear space. However many of its properties are similar to linear space. A similar concept of norm on linear space can be introduced into accounting vector space.

If Γ satisfy the following Axiom 5.2.1 then $[\Gamma]$ is called an accounting algebra. In the following section we assume that an accounting vector space satisfies the axiom of exchange.

[Definition 5.2.6] Accounting Norm
Let Λ be a set of accounting bases.

Then accounting norm $|x|$ is defined on $x \in [\Gamma]$ as follows.

$$|x| = \sum_{i=1}^{n} a_i \quad i = 1, \ldots, n$$

Where $x = \sum_{i=1}^{n} a_i e_i, a_i \in N, e_i \in \Gamma$.

We have used the term norm as a meaning of standard in Chapter 3. However, a norm in this chapter is the concept abstracted from an absolute value in mathematics.

An accounting norm is different from a norm in linear space, where triangular inequality does not hold.

[Proposition 5.2.1]

$$|x + y| = |x| + |y|$$

Proof:
It is clear from the definition. Q.E.D.

An accounting norm has a different property from the norm of linear space. This property is important for analyzing the properties of a balance sheet.

Economic transactions that are described by a double-entry bookkeeping system can also be expressed by an accounting vector. Where the concept of liability, for example, is expressed by an accounting vector without using negative numbers. Each transaction is expressed by a pair of accounting vectors satisfying a certain property. The total change of stock in a period is expressed by the sum total of the accounting vectors used for business transaction in the period. A balance sheet and profit and loss statement are also expressed by an accounting vector.

We distinguish three levels of description. The first is a transaction description by accounting algebra. The second is flow description where the accounting vector is called the accounting flow vector. The last is a state description of the accounting vector known as the accounting stock vector. Each transaction, the sum total of transactions as a flow variable in a period and the stock state of an economic agent can be expressed by an accounting vector.

[Definition 5.2.7] Simple Accounting Vector Space
Let Γ be a set of extended bases of a set of accounting bases Λ. Then a set generated by a ∈ Γ is defined as V[a] = [{a}]. This is called simple accounting vector space generated by a ∈ Γ. x ∈ V[a] is called a simple accounting vector.

The binary relation called exchange relations is defined on Γ in order to formulate a transaction by using an accounting vector. By using this binary relation, a pair of simple accounting vectors with the same norm and two bases of the pair vectors satisfy exchange relations as follows.

For example, the transaction that an apple of 500 yen is sold by cash is expressed as "500<cash> ⇔ 500 ^<apple>".

Let "<name>" be an account name then "^<name>" means the decrease of "<name>". Thus ^<apple> means the decrease of <apple> that is the transaction that an apple is sold.

Let <name> = <cash> then V[<cash>] = [{<cash>}] and, for example, 800<cash> means cash 800 dollars when we assume a dollar as the standard currency.

We introduce the following six axioms for binary relation ⇔ on Γ.

These axioms characterize what kind of transaction is significant

[Axiom 5.2.1] Axiom of Exchange

(1) $\forall\, a, b \quad a \Leftrightarrow b \equiv {}^\wedge a \Leftrightarrow {}^\wedge b$
(2) $\forall\, a, b, c \quad a \Leftrightarrow b \text{ and } b \Leftrightarrow c \rightarrow \neg(a \Leftrightarrow c)$
(3) $\forall\, a, b \quad a \Leftrightarrow b \equiv b \Leftrightarrow a$
(4) $\forall\, a, b \quad a \Leftrightarrow b \rightarrow \neg(a \Leftrightarrow {}^\wedge b)$
(5) $\forall\, a, b, c \quad \neg(a \Leftrightarrow b) \text{ and } \neg(b \Leftrightarrow c) \rightarrow \neg(a \Leftrightarrow c)$
(6) $\forall\, a\, \exists\, b \quad a \Leftrightarrow b$

Where $a, b, c \in \Gamma$ and $\neg$ means logical symbol of negation.

Let a = <cash>, b = ^<apple>, and c = <account receivable>.

Axiom (1) means that if the transaction "an apple is sold to acquire cash" is significant and acceptable (^<apple> ⇔ <cash>) then the transaction "cash is paid to purchase an apple" is also significant and acceptable (^<cash> ⇔ <apple>).

Axiom (3) means reflective law.

Axiom (2) means that if the transaction "an apple is sold to acquire cash" satisfies the exchange relation (^<apple> ⇔ <cash>) and the transaction "an apple is sold and receive accounts receivable" also satisfy the relation (^<apple> ⇔ <accounts receivable>) then the transaction to acquire cash and accounts receivable does not satisfy the relation (¬<cash> ⇔ <accounts receivable>).

Axiom (4) insists that if the transaction "an apple is sold to acquire cash" satisfies the exchange relation (^<apple> ⇔ <cash>) then the transaction to acquire cash and apple at the same time does not satisfy the relation (¬<apple> ⇔ <cash>).

However, we do not insist that it is impossible to describe the transaction "the cash and apple are obtained at the same time as a free gift" by exchange relation.

The economic activity of "the cash and apple are obtained at the same time as a free gift" is described as the transaction of the occurrence of profit by double-entry bookkeeping.

It is described as <cash> ⇔ <profit> and <apple> ⇔ <profit> in our frame work.

Axiom (5) insists that the no exchange relation (¬⇔) satisfies transitive law.

For example if there is no exchange relation between <cash> and <apple> (¬<cash> ⇔ <apple>) and there is no exchange relation between <apple> and <accounts receivable> (¬<apple> ⇔ <accounts receivable>), then the axiom insists that there is no exchange relation between <cash> and <accounts receivable> (¬<cash> ⇔ <accounts receivable>).

Axiom (6) insists that for any account name "x" there exists a certain account name "y" in Γ and "x" and "y" satisfy the exchange relation.

As a result these six axioms characterize the significant and acceptable transaction. We formulate some concepts about this transaction using the following transaction relation.

[Definition 5.2.8] Basic Transaction

When accounting vector $x, y \in [\Gamma]$ satisfies the following conditions, $\{x, y\}$ is called a basic transaction.

There exist elements of extended bases $a, b \in \Gamma$ and

(1) x, y are a simple accounting vector, that is to say $x \in V[a]$, $y \in V[b]$,
(2) a and b satisfy exchange relation as $a \Leftrightarrow b$,
(3) x and y is equal as an accounting norm $|\ |$, that is to say $|x| = |y|$.

A concept of a basic transaction gives a description of a transaction decomposed into two simple accounts such as {100^<apple>, 100<cash>}.

Next we formulate the description of a transaction in which more than two account names are used in the transaction.

[Definition 5.2.9] Transaction Pair
When accounting vector $x, y \in [\Gamma]$ satisfies the following conditions, {x, y} is called a transaction pair.

(1) $\exists n\ x = \sum_{i=1}^{n} p_i$, and $p_i \in V[L_i]$, $L_i \in \Gamma$, $i = 1, \ldots, n$.

(2) $\exists m\ y = \sum_{j=1}^{m} q_j$, and $q_j \in V[M_j]$, $M_j \in \Gamma$, $j = 1, \ldots, m$.

(3) $\forall i,j\ L_i \Leftrightarrow M_j$ and $|x| = |y|$, $i = 1, \ldots, n$, $j = 1, \ldots, m$.

[Example 5.2.3]
Let x = 50<profit> + 100^<apple> and y = 150<cash>.

{x, y} is a transaction pair.

Where p_1 = 50<profit>, p_2 = 100<apple>, q_1 = 150<cash>, n = 2, m = 1, L_1 = <profit>, L_2 = <apple>, M_1 = <cash>, <profit> ⇔ <cash> and ^<apple> ⇔ <cash> hold.

This transaction is expressed by the usual double-entry bookkeeping system as follows.

Debit side		Credit side	
Cash	150	Apple	100
		Profit	50

[Definition 5.2.10] Transaction Pair Set
A set consisting of transaction pairs is called a transaction pair set.

A transaction pair set is defined as a set which gathers all transaction pairs in a certain period.

[Definition 5.2.11] Transaction Set
Transaction set K is defined as a set of accounting vectors that consist of the transaction pair set L as follows.

$K = \{z \mid z \in \{x, y\} \text{ and } \{x, y\} \in L\}$.

Next we define a transaction flow vector as follows.

[Definition 5.2.12] Transaction Flow Vector
We define a transaction flow vector V[K] as

$$V[K] = \sum\{x \mid x \in K\}.$$

Where K is a transaction set.

A transaction flow vector is introduced from a suitable transaction set as a summation of the accounting vectors of K.

This summation of vectors corresponds to the calculation of the total balance sheet. Under these preparations we show that properties of double-entry bookkeeping can be derived from the axioms of accounting vector space.

We give the relation between accounting vectors and double-entry bookkeeping. Various properties of double-entry bookkeeping, for example, the balance of the right and left side of a balance sheet, are shown as a theorem deduced from the axiom of exchange.

At first we introduce a relation called transaction impossible relation on an extended base Γ. This relation becomes an equivalence relation and Γ is divided into two equivalence classes. It is very important that there exist only two equivalence classes. One of these equivalence classes is named the debit side, and the other is named the credit side.

[Definition 5.2.13] Transaction Impossible Relation ▲
For arbitrary extended bases a, b ∈ Γ if $\neg(a \Leftrightarrow b)$ holds then "a" and "b" are called transaction impossible and described as a ▲ b.

Binary relation ▲ on Γ is called a transaction impossibility relation.

[Theorem 5.2.1]
Transaction impossible relation ▲ is an equivalence relation on Γ and Γ is divided into two equivalence classes by this equivalent relation.

Proof:
At first we show that transaction impossibility relation is an equivalence relation on Γ.

(1) Reflexivity
We show a ▲ a for any a ∈ Γ, that is $\neg(a \Leftrightarrow a)$.
From axiom (6) $\forall a \, \exists b \; a \Leftrightarrow b$. $b \Leftrightarrow a$ holds from axiom (3). Then $a \Leftrightarrow b$ and $b \Leftrightarrow a \rightarrow \neg(a \Leftrightarrow a)$ holds from axiom (2).

(2) Symmetry
We show a ▲ b → b ▲ a for any a, b ∈ Γ, that is $\neg(a \Leftrightarrow b) \rightarrow \neg(b \Leftrightarrow a)$. It is clear from the axiom (3).

(3) Transitivity
We show a ▲ b and b ▲ c → a ▲ c for any a, b, c ∈ Γ, that is $\neg(a \Leftrightarrow b)$ and $\neg(b \Leftrightarrow c) \rightarrow \neg(a \Leftrightarrow c)$. It is clear from the axiom (5).

Next we show that there are only two equivalence classes on Γ.

For this purpose it is enough to show that $\neg(a \blacktriangle b)$ and $\neg(a \blacktriangle c) \rightarrow b \blacktriangle c$ for any $a, b, c \in \Gamma$, that is b and c belong to the same equivalence class.

From the definition of a transaction impossibility relation, it is shown that $(a \Leftrightarrow b)$ and $(a \Leftrightarrow c) \rightarrow \neg(b \Leftrightarrow c)$ for any $a, b, c \in \Gamma$. It is clear from axiom (3) and (2). Q.E.D.

As a result we introduced two equivalence classes on the Γ from the transaction impossibility relation. This is why double-entry bookkeeping has a debit and credit side for this description.

We formulate the debit side and credit side as an equivalence class on Γ as follows.

[Definition 5.2.14]
$\Gamma/\blacktriangle$ = {debit side, credit side} is called a set of accounting equivalence classes.

debit side = {a | a∈Γ, ∃c ∈Λ a ▲ c, k(c) = <assets> ∈ Ω}.

credit side = {a | a∈Γ, ∃c ∈Λ a ▲ c and k(c) = <liability> ∈ Ω}.

Where $k: \Lambda \rightarrow \Omega$ is a function which classifies the set of accounting bases, defined at Definition 5.2.4.

If "a" belongs to the debit side then "^a" belongs to the credit side and that if "a" belongs to the credit side then "^a" belongs to the debit side contrary to the Definition and Theorem 5.2.1.

In general the set of extended bases Γ is decomposed from two different points of view as follows.

(1) $\Gamma = \Lambda \cup {}^\wedge\Lambda$ and $\Lambda \cap {}^\wedge\Lambda = \phi$
(2) Γ = debit side ∪ credit side and debit side ∩ credit side = ϕ.

It is important that the two kinds of decomposition of Γ are not equal ones.

[Example 5.2.4]
Let Λ = {<assets>, <liability>, <equity>, <profit>, <cost>} and ^Λ = {^<assets>, ^<liability>, ^<equity>, ^<profit>, ^<cost>}.

Then the debit side and the credit side are defined as follows.

debit side = {<assets>, ^<liability>, ^<equity>, ^<profit>, <cost>}

credit side = {^<assets>, <liability>, <equity>, <profit>, ^<cost>} Γ = Λ ∪ Λ = debit side ∪ creditor

Where we assume that the exchange relation is given depending on the standard interpretation of double-entry bookkeeping.

[Definition 5.2.15] Debit–Side Vector, Credit–Side Vector
[debit side] denotes an accounting vector space that generated from bases belonging to debit side. [credit side] denotes an accounting vector space that generated from bases belonging to crebit side.

x is called debit–side vector if $x \in$ [debit side] and called credit–side vector if $x \in$ [credit side].

[Definition 5.2.16] Reduction Map
Reduction map $f_{rd}[\Gamma] \to [\Lambda]$ is defined as follows.

(1) $f_{rd}(ae) = ae$ if $e \in \Lambda$
$= a\hat{\ }e$ if $e \in \hat{\ }\Lambda$
(2) $f_{rd}(x + y) = f_{rd}(x) + f_{rd}(y)$, $f_{rd}(ax) = af_{rd}(x)$, where $x,y \in [\Gamma]$ and $a \in N$.

[Proposition 5.2.2]
Reduction map becomes one to one map if the domain of map is $[\Lambda]$, $[\hat{\ }\Lambda]$, [debit side] or [credit side].

Proof:
Let A be a subset of extended base Γ. We also assume that if $e \in A$ then $\neg(\hat{\ }e \in A)$. Then reduction map on [A] becomes one to one map from the definition of reduction map. $[\Lambda]$, $[\hat{\ }\Lambda]$, [debit side] and [credit side] satisfy this property. Q.E.D.

[Definition 5.2.17] Expansion Map
f_{exr}: $[\Lambda] \to [\Gamma]$ and f_{exl}: $[\Lambda] \to [\Gamma]$ are called expansion map if they satisfy the following conditions.

(1) $f_{exl}(x + y) = f_{exl}(x) + f_{exl}(y)$
$f_{exl}(ax) = a\, f_{exl}(x)$, where $x,y \in [\Lambda]$ and $a \in N$

f_{exr} also satisfy the same property.

(2A) $f_{exl}(ae) = ae$, if $e \in$ debit side,
$= a\hat{\ }e$, if $e \in$ credit side.
(2B) $f_{exr}(ae) = ae$, if $e \in$ credit side,
$f_{exr}(ae) = a\hat{\ }e$, if $e \in$ debit side.

[Proposition 5.2.3]
An accounting vector $x \in [\Gamma]$ is decomposed uniquely as the summation of a debit–side vector and credit–side vector. That is to say, symbolically, $\forall x \in [\Gamma] \exists!\ a \in$ [debit side] $\exists!\ b \in$ [credit side] $x = a + b$

Proof:
It is clear from the definition of accounting vector. Q.E.D.

[Proposition 5.2.4]
Either of the next two statement is true for a transaction pair {x,y}.

(1) x is a credit–side vector and y is a debit side vector.
(2) x is a debit side vector and y is a credit–side vector.

Proof:
Let {x,y} be a transaction. Suppose that

$$x = \sum_{i=1}^{n} p_i, \quad y = \sum_{j=1}^{m} q_j,$$

hold for some $p_i \in V[L_i]$, some $q_j \in V[M_j]$, m and n.

Suppose $L_i \Leftrightarrow M_j$ for any $i = 1, \ldots, n, j = 1, \ldots, m$.

Then we have either of the following.

(1) If $L_i \in$ debit side for some i, then $L_i \in$ debit side for any i and $M_j \in$ credit side for any j.
(2) If $L_i \in$ credit side for some i, then $L_i \in$ credit side for any i and $M \in$ debit side for any j. Q.E.D.

[Definition 5.2.18] Tabular Expression of Transaction
Let {x,y} be a transaction.

(1) Let x be a debit–side vector any y be a credit–side vector, then the ordered pair $<f_{rd}(x),f_{rd}(y)>$ is called the tabular expression of {x,y}.
(2) Let x be a credit–side vector and y be a debit–side vector, then the ordered pair $<f_{rd}(y),f_{rd}(x)>$ is called the tabular expression of {x,y}.

Where f_{rd}: $[\Gamma] \rightarrow [\Lambda]$ is the reduction map that is defined above.

g({x,y}) denotes the tabular expression of {x,y} as follows, that is called the tabular formation function.

(1) $g(\{x,y\}) = <f_{rd}(x),f_{rd}(y)>$, if x be a debit–side vector and y be a credit–side vector.
(2) $g(\{x,y\}) = <f_{rd}(y),f_{rd}(x)>$, if x be a credit–side vector and y be a debit–side vector.

We do not need to use the set of dual account bases $^\wedge\Lambda$ in tabular expression, because the accounting vector space is reduced from $[\Gamma]$ to $[\Lambda]$. Instead of unordered pair {x,y} we must use ordered pair $<f_{rd}(x),f_{rd}(y)>$ or $<f_{rd}(y),f_{rd}(x)>$.

Next we show that the tabular expression of a transaction is equivalent to a transaction expressed by the unordered pair.

[Definition 5.2.19] Tabular Expression of Transaction Pair Set
Let L be a transaction pair set. Then the set $M = \{<x,y> \mid <x,y> = g(\{a,b\}), \{a,b\} \in L\}$ is called the tabular expression transaction pair set of L.

Where g({x,y}) is the tabular formation function.

[Proposition 5.2.5]
Let L be a transaction pair set.
Let M be its tabular expression.
Then the tabular formation function $g: L \rightarrow M$ is one to one map (bijection).

Proof:
It is clear from definition. Q.E.D.

Inverse function of this tabular formation function g^{-1} is defined as follows.

$$g^{-1}(<x,y>) = \{f_{exl}(x), f_{exr}(y)\}$$

Where f_{exl}: $[\Lambda] \rightarrow [\Gamma]$ and f_{exr}: $[\Lambda] \rightarrow [\Gamma]$ are expansion map. Consequently the set L and M are isomorphic and we can use either description. We describe a transaction as the tabular form in usual bookkeeping system. However the tabular description is rather inconvenient for the theoretical treatment. For example it is impossible for a tabular form to express the sum total of transaction in a given period as one accounting vector.

[Example 5.2.5]
Let x = 500^<Cash> and y = 500<Apple>. Then the transaction in which we buy 500 yen worth of apples for cash can be expressed by

{x,y} = {500<Apple>,500^<Cash>},
g({x,y}) = <500<Apple>,500<Cash>> and
g^{-1}(<500<Apple>,500<Cash>>) = {500<Apple>,500^<Cash>}.

In the usual double entry bookkeeping system, an ordered pair in the form of a tabular description is used. Finally we will show the well known property of equality between the debit side and the credit side by our framework.

[Theorem 5.2.2]
Let K be a transaction set. We assume $KR \cap KL = \phi$, $KR \cup KL = K$, $KL \subseteq$ [debit side], $KR \subseteq$ [credit side].

Then KR and KL are determined uniquely. Moreover, if

$$y_R = \sum_{x \in KR} x \text{ and } y_L = \sum_{x \in KL} y$$

hold then we have $|y_R| = |y_L|$.

Proof:
From the definition of transaction set, we have

$$K = \bigcup_{i=1}^{n} A_i, \quad A_i = \{a_i, b_i\},$$

where for all i ($i = 1, \ldots, n$), $A_i = \{a_i, b_i\}$ and $|a_i| = |b_i|$ hold, a_i is a credit–side vector and b_i is a debit–side vector. Let

$KR = \{a \mid a$ is a credit–side vector and $a \in A_i$ for some $i\}$
$KL = \{b \mid b$ is a debit–side vector and $b \in A_i$ for some $i\}$

Then we have

$$KL \cup KR = \bigcup_{i=1}^{n} A_i = K \text{ and } KL \cap KR = \phi.$$

Uniqueness of KL and KR is clear from its construction. Moreover we have

$$|y_R| = \left|\sum_{i=1}^{n} a_i\right| = \sum_{i=1}^{n} |a_i|,$$

where $a_i \in A_i$ and a_i is a credit–side vector, and

$$|y_L| = \left|\sum_{i=1}^{n} b_i\right| = \sum_{i=1}^{n} |b_i|,$$

where $b_i \in A_i$ and b_i is a debit–side vector.

Since $|a_i| = |b_i|$ for each i, we have $|y_R| = |y_L|$. Q.E.D.

We can express the above theorem more directly in the tabular expression.

Let L be a transaction pair set.

Let

$M_L = \{x \mid <x,y> = g(\{a,b\}) \text{ and } \{a,b\} \in L\}$,
$M_R = \{y \mid <x,y> = g(\{a,b\}) \text{ and } \{a,b\} \in L\}$.

Let $a_R = \sum_{y \in M_R} y, a_L = \sum_{x \in M_L} x$.

Then we have $|a_R| = |a_L|$.

Next we make clear the relation between double entry bookkeeping system as an accounting vector space and single entry bookkeeping system described as a Z–additive group of integers. Finally we will introduce an equivalence relation ~ on an accounting vector space B, and show that B/~ is isomorphic to the Z–additive group A. This ties together the single entry description and the double entry description.

[Definition 5.2.20]
Let A be a Z-additive group generated by Λ and B be an accounting vector space $[\Gamma]$. Then function $f: A \to B$ is defined as follows.

$$f(x) = \sum_{a_i \geqq 0} a_i e_i + \sum_{a_i < 0} |a_i|{}^\wedge e_i, \text{ where } x = \sum_{e_i \in \Lambda} a_i e_i \in A$$

The function is called double-entry function.

[Proposition 5.2.6]
Double-entry function $f: A \to B$ is a one to one map but is not an onto map.

Proof:
One to one is clear from definition. We give a counter example to show f is not an onto map. Let $y = e_1 + 2{}^\wedge e_1$. Then there is no "x" such as $f(x) = y$.

Because x dose not include $1e_1$ and $-2e_1$ at the same time from the definition of A. Q.E.D.

[Proposition 5.2.7]
Let f be a double-entry function then the following equation holds.
$\exists x,y \in A \; f(x + y) \neq f(x) + f(y)$

Proof:
Let $x = 3e_1$, $y = -2e_1 + 4e_2$. Then

$f(x + y) = f(e_1 + 4e_2) = e_1 + 4e_2$,
$f(x) + f(y) = f(3e_1) + f(-2e_1 + 4e_2) = 3e_1 + 2{}^\wedge e_1 + 4e_2$.

Therefore $f(x + y) \neq f(x) + f(y)$ holds. Q.E.D.

We introduce the equivalence relation such that $3e_1 + 2{}^\wedge e_1$ is identified with e_1.

[Definition 5.2.21] Standard Elements
Let B be an accounting vector space [Γ] and $x \in B$.

Then x is expressed as $x = \sum_{e_i \in \Lambda} a_i e_i + \sum_{e_i \in \Lambda} b_i {}^\wedge e_i$.

Then standard element $\overline{}x$ of x is defined as follows.

$$\overline{}x = \sum_{e_i \in A} |a_i - b_i| \, g(a_i, b_i, e_i)$$

Where $g(a_i, b_i, e_i) = e_i$ if $a_i - b_i \geqq 0$, $g(a_i, b_i, e_i) = {}^\wedge e_i$ if $a_i - b_i < 0$.

[Definition 5.2.22] Equivalence Relation "~" on B
We define accounting equivalence relation "~" on accounting vector space B as follows.

$x \sim y \equiv \overline{}x = \overline{}y,\ x \in B$

[Proposition 5.2.8]
The relation "~" is an equivalence relation on B.

Proof:
We show (1) $x \sim x$, (2) $x \sim y \to y \sim x$ and (3) $x \sim y$ and $y \sim x \to x \sim y$.
It is clear from the definition. Q.E.D.

$\overline{}x$ becomes a representative of equivalence class $[x] \in B/\sim$.

[Definition 5.2.23] Operation on Quotient Set B/~
We introduce the operation on the quotient set B/~ as follows.

(1) $[x] + [y] = [\overline{}x + \overline{}y]$
(2) $a[x] = [a\overline{}x]$ for $a \in N$

[Theorem 5.2.3]
The set of standard elements on accounting vector space B becomes isomorphic to Z-additive group A as sets.

Proof:
Let $\overline{}B = \{\overline{}Y \mid y \in B\}$, $h(x) = \overline{}x$.
m: $A \to B$ is defined as $m(x) = h(f(x))$, where f is the double-entry function. Then we show "m" is one to one and onto map. One to one is clear from the definition of h(f(x)). Thus we show that $\forall y \in \overline{}B\ \exists x \in A\ h(f(x)) = y$.
Let $y \in \overline{}B$ then y is described as follows.

$$y = \sum_{e_i \in S} c_i e_i + \sum_{e_j \in T} d_j {}^\wedge e_j, \text{ where } S \cap T = \phi, S \cup T = \Lambda, c_i, d_i \geqq 0$$

$x \in A$ is defined as $x = \sum_{e_i \in S} c_i e_i - \sum_{e_j \in T} d_j e_j$,

then $m(x) = y$ holds. Q.E.D.

[Proposition 5.2.9]
Relation "~" is a congruence relation.

Proof:
We show that $[ax + by] = a[x] + b[y]$. Thus we show
(1) $[ax] = a[x]$ and
(2) $[x + y] = [x] + [y]$.
$[ax] = a[x]$ is clear form $h(ax) = ah(x)$.
Where $h(x) = {}^{-}x$.
$[x] + [y] = [{}^{-}x + {}^{-}y]$ from the definition.
We show $h(x + y) = h(h(x) + h(y))$ for $[{}^{-}x + {}^{-}y] = [x + y]$.

$$\text{Let } x = \sum_{e_i \in \Lambda} a_i e_i + \sum_{e_i \in \Lambda} b_i \hat{\ } e_i, y = \sum_{e_i \in \Lambda} c_i e_i + \sum_{e_i \in \Lambda} d_i \hat{\ } e_i,$$

$$h(x+y) = \sum_{e_i \in \Lambda} |a_i - b_i + c_i - d_i| \, g(a_i + c_i, b_i + d_i, e_i)$$

$$= h\left(\sum_{e_i \in \Lambda} |a_i - b_i| \, g(a_i, b_i, e_i) + \sum_{e_i \in \Lambda} |c_i - d_i| \, g(c_i, d_i, e_i) \right)$$

$$= h(h(x) + h(y))$$

Q.E.D.

The calculation of standard lement from accounting flow vector "x" means the calculation of the trial balance sheet.

[Theorem 5.2.4]
Let $F: A \to B/\sim$ and $F(x) = [f(x)]$, where f is the double-entry function.
Then F is an isomorphic map from A to $B/\sim$ as Z-additive group.

Proof:

$$\text{Let } x = \sum_{e_i \in \Lambda} a_i e_i, y = \sum_{e_i \in \Lambda} b_i e_i, a_i b_i \in Z.$$

$$F(x) = [f(x)] = \left[\sum_{a_i \geqq 0} a_i e_i + \sum_{a_i < 0} |a_i| \hat{\ } e_i \right],$$

We show (1) $F(x + y) = F(x) + F(y)$,
(2) $F(cx) = cF(x)$ if $c \geqq 0$, $F(cx) = |c| \, F(-x)$ if $c < 0$.

(1) $F(x + y) = F(x) + F(y)$
$\equiv [f(x + y)] = [f(x)] + [f(y)]$
$\equiv h(f(x + y)) = h(h(f(x)) + h(f(y)))$.

$$f(x) = \sum_{e_i \in \Lambda} c_i e_i + \sum_{e_i \in \Lambda} d_i \hat{\ } e_i,$$

where if $a_i \geqq 0$ then $c_i = a_i$, $d_i = 0$,
if $a_i < 0$ then $c_i = 0$, $d_i = -a_i$.

$$f(y) = \sum_{e_i \in \Lambda} v_i e_i + \sum_{e_i \in \Lambda} w_i e_i,$$

where if $b_i \geqq 0$ then $v_i = b_i$, $w_i = 0$,
if $b_i < 0$ then $v_i = 0$, $w_i = -b_i$.
Then $h(f(x)) = f(x)$.

$$h(f(x+y)) = f(x+y) = \sum_{a_i+b_i \geqq 0} (a_i + b_i)e_i + \sum_{a_i+b_i<0} -(a_i + b_i)\hat{}e_i.$$

$$h(h(f(x)) + h(f(y))) = h(f(x) + f(y))$$

$$= h\left(\sum_{e_i \in \Lambda} (c_i + v_i)e_i + \sum_{e_i \in \Lambda} (d_i + w_i)\hat{}e_i\right)$$

$$= \sum_{e_i \in \Lambda} |c_i + v_i - (d_i + w_i)|\, g(c_i + v_i, d_i + w_i, e_i)$$

$$= \sum_{a_i+b_i \geqq 0} (a_i + b_i)e_i + \sum_{a_i+b_i<0} -(a_i + b_i)\hat{}e_i.$$

Thus $F(x + y) = F(x) + F(y)$ holds.

(2) $F(cx) = cF(x)$ if $c \geqq 0$, $F(cx) = |c|\,F(-x)$ if $c < 0$
$\equiv [f(cx)] = c[f(x)]$ if $c \geqq 0$, $[f(cx)] = |c|\,[f(-x)]$ if $c < 0$
$\equiv h(f(cx)) = ch(f(x))$ if $c \geqq 0$, $h(f(cx)) = |c|\,h(f(-x))$ if $c < 0$.

$$f(x) = \sum_{a_i \geqq 0} a_i e_i + \sum_{a_i<0} |a_i|\hat{}e_i.$$

If $c \geqq 0$ then

$$f(cx) = \sum_{a_i \geqq 0} ca_i e_i + \sum_{a_i<0} c\,|a_i|\hat{}e_i.$$

Then $h(f(cx)) = h(cf(x)) = ch(f(x))$ holds.

If $c < 0$ then

$$f(cx) = f\left(\sum_{e_i \in \Lambda} ca_i e_i\right) = \sum_{a_i \geqq 0} |ca_i|\hat{}e_i + \sum_{a_i<0} ca_i e_i, \text{ where } ca_i \geqq 0 \text{ if } a_i < 0, c < 0,$$

$$= |c| \sum_{a_i \geqq 0} a_i\hat{}e_i + |c| \sum_{a_i<0} |a_i|\,e_i$$

$$= |c|\left(\sum_{-a_i \geqq 0} |a_i|\,e_i + \sum_{-a_i<0} a_i\hat{}e\right)$$

$$= |c|\left(\sum_{-a_i \geqq 0} -a_i e_i + \sum_{-a_i<0} |-a_i|\hat{}e_i\right)$$

$$= |c|\,f(-x).$$

Then $h(f(cx)) = h(|c|\,f(-x)) = |c|\,h(f(-x))$ holds.

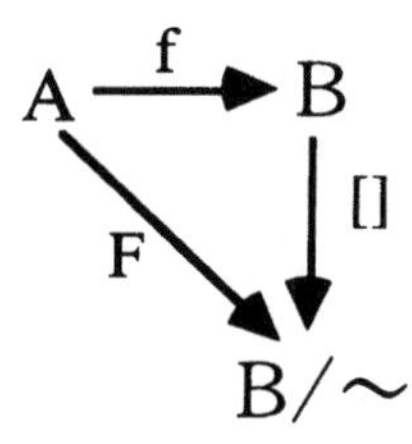

FIG. 5.1 Commutative Map

Thus F becomes homomorphism from A to B/~. We have already shown that F is isomorphism as sets. Consequently we have the results. Q.E.D.

We obtain the following commutative diagram.

5.3 Exchange Algebra

5.3.1 Concept of Exchange and Economics

The concept of exchange in economic systems is usually used for analyzing the structure of a market. In this paper, however, we study the concepts of exchange as a descriptive basis of economic systems and investigate their use for analyzing the common structure of different economic systems.

5.3.2 Redundant Algebra

We previously introduced the concept of accounting vector space as an abstraction of the bookkeeping system [Deguchi, 1986]. Now we give a further abstraction and give basic concepts to develop economic analysis with the algebra.

At first we introduce the notion of redundant algebra corresponding to the vector spaces used in natural sciences. Redundant algebra does not use the notion of negative elements. Instead of using negative elements we add redundant elements to the algebra by using the concept of a dual basis. In order to formulate the notion of exchange in economics we need to add several axioms to the algebra. Then we can formulate the mathematical structure of exchange, which has been ambiguous in economics.

[Definition 5.3.1] Redundant Algebra Ψ
Let Ψ be a non-empty set and T be a set of non-negative integers Z^+ or non-negative real numbers R^+. We call T the set of scalars. We define a binary operation + and unary operations ^, $\overline{\ }$ on Ψ and a scalar multiplication $ax \in \Psi$ for $a \in T$, $x \in \Psi$, which satisfy the following axioms.

(1) $x + y = y + x$
(2) $(x + y) + z = x + (y + z)$
(3) $x + 0 = x$
(4) $a(bx) = (ab)x$
(5) $1x = x, 0x = 0$
(6) $(a + b)x = ax + bx$
(7) $a(x + y) = ax + ay$
(8) $\overline{\ }\,\overline{\ }x = \overline{\ }x$
(9) $\hat{}\hat{}x = x$
(10) $\overline{\ }(x + y) = \overline{\ }(\overline{\ }x + \overline{\ }y)$
(11) $\hat{}(x + y) = \hat{}x + \hat{}y$
(12) $\overline{\ }(x + \hat{}y) = 0 \equiv \overline{\ }x = \overline{\ }y$
(13) $\overline{\ }(\hat{}x) = \hat{}(\overline{\ }x)$
(14) $\overline{\ }(ax) = a(\overline{\ }x), \hat{}(ax) = a(\hat{}x)$
(15) $x + y = 0 \rightarrow x = 0$ and $y = 0$
(16) $x + z = x \rightarrow z = 0$
(17) $ax = 0 \rightarrow a = 0$ or $x = 0$

Where $a, b \in T$ and $x, y, z \in \Psi$.

Then we call Ψ a redundant algebra or a redundant space. We speak of real redundant algebra or a Z^+-redundant algebra according to $T = R^+$ or $T = Z^+$ respectively.

[Proposition 5.3.1]
(1) $\overline{\ }(x + \hat{}x) = 0$
(2) $\hat{}0 = 0, \overline{\ }0 = 0$
(3) $\hat{}x = 0 \equiv x = 0$

Proof:
(1) From axiom (12) we get $\overline{\ }(x + \hat{}x) = 0 \equiv \overline{\ }x = \overline{\ }x$ and $\overline{\ }x = \overline{\ }x$ is clear.
(2) From $x = x + 0$, $\hat{}x = \hat{}x + \hat{}0$. From axiom (16), $\hat{}0 = 0$ holds. Thus $\hat{}0 + 0 = 0 + 0 = 0$.

Then $\overline{\ }(\hat{}0 + 0) = \overline{\ }0$ and $\overline{\ }(\hat{}0 + 0) = 0$ from (1) of this proposition. Consequently $\overline{\ }0 = 0$ holds.

(3) Let $\hat{}x = 0$. Then $x + \hat{}x = x + 0 = x$ and $\hat{}(x + \hat{}x) = \hat{}x + x = \hat{}x$. From axiom (16) $x = 0$ holds. Conversely we assume $x = 0$. $\hat{}x = \hat{}(x + 0) = \hat{}x + \hat{}0$. $\hat{}x = \hat{}0$ from assumption.

Thus $\hat{}x = \hat{}x + \hat{}0 = \hat{}0 + \hat{}0 = \hat{}(0 + 0) = \hat{}0$, From axiom (16) $\hat{}x = 0$ holds.
Q.E.D.

[Example 5.3.1] Redundant Real Numbers and Redundant Sequences
(a) Redundant Real Numbers
Let $R_r = \{(x_1, x_2) \mid x_1, x_2 \in R^+\}$.
The operations on R_r are given as follows.

(1) $(x_1, x_2) + (y_1, y_2) = (x_1 + y_1, x_2 + y_2)$
(2) $a(x_1, x_2) = (ax_1, ax_2)$
(3) $\hat{}(x_1, x_2) = (x_2, x_1)$
(4) $\overline{\ }(x_1, x_2) = (x_1 - x_2, 0)$ if $x_1 \geq x_2$,
$= (0, x_2 - x_1)$ if $x_2 > x_1$.

Then R_r satisfies the axioms of redundant algebra and R_r is called a set of redundant real numbers.

(b) Redundant Sequences
Let $l_{nr} = \{x \mid x = ((x_{11}, x_{12}), (x_{21}, x_{22}), \ldots, (x_{n1}, x_{n2})) \mid x_{ij} \in T)$.
The operations on l_{nr} are given as follows.

(1) $x + y = ((x_{11} + y_{11}, x_{12} + y_{12}), \ldots, (x_{n1} + y_{n1}, x_{n2} + y_{n2}))$
(2) $ax = ((ax_{11}, ax_{12}), (ax_{21}, ax_{22}), \ldots, (ax_{n1}, ax_{n2})), a \in T)$.
(3) $\hat{}x = ((x_{12}, x_{11}), (x_{22}, x_{21}), \ldots, (x_{n2}, x_{n1})$
(4) $\overline{\ }x = ((c_{11}, c_{12}), (c_{21}, c_{22}), \ldots, (c_{n1}, c_{n2}))$

where $c_{i1} = x_{i1} - x_{i2}, c_{i2} = 0$ if $x_{i1} \geq x_{i2}$,
$c_{i2} = x_{i2} - x_{i1}, c_{i1} = 0$ if $x_{i2} > x_{i1}$.

Then l_{nr} satisfies the axioms of redundant algebra and l_{nr} is called a set of redundant sequences.

[Definition 5.3.2] Subspace
A non-empty subset M of a redundant space Ψ is called a redundant subspace of Ψ, if M satisfies the following conditions.

(1) if $x, y \in M$ then $x + y \in M$ (2) $\forall x \in M, a \in T$ $ax \in M$
(3) $\forall x \in M$ $\hat{}x \in M$ (4) $\forall x \in M$ $\bar{}x \in M$

If M satisfies only the condition (1), (2), then M is called a weak redundant subspace of Ψ. Next we introduce a similar notion of the norm of usual linear spaces.

[Definition 5.3.3] Redundant Norm
A norm on a redundant space Ψ is a function which assigns a real number $|x|$ to each element x in the Ψ in such a manner that

(1) $|x| \geqq 0$ and $|x| = 0 \equiv x = 0$ (2) $|ax| = a|x|$
(3) $|x + y| \leqq |x| + |y|$ (4) $|\hat{}x| = |x|$
(5) $|\bar{}x| \leqq |x|$ and $|\bar{}x| = |x| \equiv \bar{}x = x$

[Example 5.3.2] Redundant Norm
(1) Let $R_r = \{(x_1, x_2) \mid x_1, x_2 \in R^+\}$ be the set of redundant real numbers. Then $|(x_1, x_2)| = x_1 + x_2$ becomes the redundant norm.
(2) Let $l_{nr} = \{x \mid x = ((x_{11}, x_{12}), (x_{21}, x_{22}), \ldots, (x_{n1}, x_{n2})) \mid x_{ij} \in T\}$ be a set of redundant sequences. Then we can define the redundant norm on a set of redundant sequences as follows.

$$|x| = \sum_{i=1}^{n} x_{i1} + x_{i2}.$$

(3) Let l_{nr} be a set of redundant sequences. Then we can introduce L^k norm on redundant sequences as follows.

$$|x|_k = \left\{\sum_{i=1}^{n}\left\{(x_{i1})^k + (x_{i2})^k\right\}\right\}^{1/k}$$

It is easy to show that $|x + y| = |x| + |y|$ holds if $k = 1$. We use this norm as a standard norm on redundant algebra.

[Proposition 5.3.2] $|\bar{}(x + y)| \leqq |x| + |y|$

Proof:
It is clear from the axiom (5) and (3). Q.E.D.

We can introduce the notions of linear combination, linear dependence, linear independence, dimension, and basis according to linear spaces as follows [Deguchi, 1984B, 1986, 1990].

[Definition 5.3.4] Linear Combination and Weak Linear Combination
Let Ψ be a redundant space, and if $\{x_1, \ldots, x_n\}$ be a finite nonempty set of elements in Ψ, then an element $x = a_1x_1 + a_2x_2 + \ldots, a_nx_n + b_1\hat{}x_1 \ldots b_n\hat{}x_n$ is called a linear combination of $x_1, \ldots, x_n$, where $a_1, \ldots, a_n, b_1, \ldots, b_n \in T$. And $y = a_1x_1 + a_2x_2 + \ldots + a_nx_n$ is called a weak linear combination of $x_1, \ldots, x_n$, where $a_1, \ldots, a_n \in T$.

[Definition 5.3.5] Linear Independence, Linear Dependence, Dimension, Basis
Let Ψ be a redundant space, and let $S = \{x_1, \ldots, x_n\}$ be a finite non-empty set of elements in Ψ. S is said to be linearly dependent if there exists scalars $a_1, a_2, \ldots, a_n, b_1, b_2, \ldots, b_n$, for some $i(1 \leqq i \leqq n)$ $a_i \neq b_i$, such that

$$\overline{}(a_1x_1 + a_2x_2 + \ldots + a_nx_n + b_1{}^\wedge x_1 + \ldots + b_n + {}^\wedge x_n) = 0. \tag{1}$$

If S is not linearly dependent, then it is called linearly independent; and this clearly means that if equation (1) holds for certain scalar coefficients $a_1, \ldots, a_n, b_1, \ldots, b_n$, then for all these scalar $a_i = b_i$ $(1 \leqq i \leqq n)$ necessarily, i.e., if $\overline{}(a_1x_1 + a_2x_2 + \ldots + a_nx_n + b_1{}^\wedge x_1 + \ldots + b_n{}^\wedge x_n) = 0$ then $a_1 = b_1, \ldots, a_n = b_n$.

Let Ψ be a redundant space. If Ψ contains n independent elements and every $n+1$ elements always become dependent, then Ψ is called n-dimensional. If there exists n independent elements for any $n \in N$, then Ψ is called infinite-dimensional. Let Ψ be an n-dimensional redundant space, then a set of n independent elements is called a basis for Ψ.

[Proposition 5.3.3]
Let Ψ be an n-dimensional redundant space. If $e_1, e_2, \ldots, e_n$ are independent elements in Ψ, then $\forall x \in \Psi$ is uniquely expressed such as $x = a_1e_1 + \ldots + a_ne_n + b_1{}^\wedge e_1 + \ldots + b_n{}^\wedge e_n$.

Proof:

Let $x = \sum_{i=1}^{n} a_ie_i + \sum_{i=1}^{n} b_i{}^\wedge e_i, y = \sum_{i=1}^{n} a_i'e_i + \sum_{i=1}^{n} b_i'{}^\wedge e_i$, and $x = y$.

Then we get $\overline{}(x + {}^\wedge y) = 0$ from axiom (12) and $\overline{}x = \overline{}y$.

Thus $\overline{}(\sum_{i=1}^{n}(a_i + b_i')e_i + \sum_{i=1}^{n}(a_i' + b_i){}^\wedge e_i) = 0$ holds.

We assume that $e_1, e_2, \ldots, e_n$ are independent, then $a_i + b_i' = a_i' + b_i$, $i = 1, \ldots, n$ holds. Let $c_i = a_i - a_i' = b_i - b_i'$, then we show that c_i become 0 for any i.

From axiom (16) and $x = y$, $y + z = x \rightarrow z = 0$ holds. We can assume that $c_i \geqq 0$, $a_i = a_i' + c_i$, $b_i = b_i' + c_i$.

Thus we can assume $z = \sum_{i=1}^{n} c_ie_i + \sum_{i=1}^{n} c_i{}^\wedge e_i$ and $y + z = x$ and $z = 0$ hold.

From axiom (15), $c_ie_i = 0$, $c_i{}^\wedge e_i = 0$, for $i = 1, \ldots, n$ hold. Then we get $c_i = 0$ from $e_i \neq 0$, ${}^\wedge e_i \neq 0$. Q.E.D.

[Definition 5.3.6]
Let A be an arbitrary non-empty subset of an n-dimensional redundant space Ψ. We denote the smallest subspace which contains A by $[A]_r$. We denote the set of all weak linear combinations of elements in A by [A]. Usually [A] is not a redundant space.

[Definition 5.3.7] Basic Basis, Dual Basis, Extended Basis
Let Λ be a basis of Ψ. If $|x| = 1$ for $\forall x \in \Lambda$, then Λ is called a basic basis of Ψ and ${}^\wedge\Lambda = \{{}^\wedge e \mid e \in \Lambda\}$ is called the dual basis of Λ and $\Gamma = \Lambda \cup {}^\wedge\Lambda$ is called the extended basis of Ψ.

[Proposition 5.3.4]

(1) If Λ be a basis of an n-dimensional redundant space Ψ, then $\wedge\Lambda$ is a basis of Ψ.

(2) If Λ be a basis of an n-dimensional redundant space Ψ, then $[\Lambda]_r = [\wedge\Lambda]_r = [\Lambda \cup \wedge\Lambda] = \Psi$

Proof:

(1) It is clear from $\wedge\wedge x = x$ and $\wedge x \in \wedge\Lambda \equiv \wedge\wedge x = x \in \Lambda$.

(2) $[\Lambda]_r = [\wedge\Lambda]_r$ is clear from (1). $[\Lambda]_r = [\Lambda \cup \wedge\Lambda]$ is clear form Proposition 5.3.3.

Q.E.D.

5.3.3 Exchange Algebra

We add the axioms to redundant algebra and define exchange algebra.

[Definition 5.3.8] Exchange Algebra

Let Λ be a basic basis of a redundant space Ψ and $\Gamma = \Lambda \cup \wedge\Lambda$ be the extended basis of Λ. If there exist a relation $\Leftrightarrow$ on Γ satisfying the next six axioms, then Ψ is called an exchange algebra. We call the relation $\Leftrightarrow$ an exchange relation and the basic basis Λ on which the relation $\Leftrightarrow$ is defined is called an exchange basis of an exchange algebra and Γ is called an extended exchange basis.

(1) $\forall x, y \in \Gamma\ x \Leftrightarrow y \equiv \wedge x \Leftrightarrow \wedge y$

(2) $\forall x, y, z \in \Gamma\ x \Leftrightarrow y$ and $y \Leftrightarrow z \rightarrow \neg(x \Leftrightarrow z)$

(3) $\forall x, y \in \Gamma\ x \Leftrightarrow y \equiv y \Leftrightarrow x$

(4) $\forall x, y \in \Gamma\ x \Leftrightarrow y \rightarrow \neg(x \Leftrightarrow \wedge y)$

(5) $\forall x, y, z \in \Gamma\ \neg(x \Leftrightarrow y)$ and $\neg(y \Leftrightarrow z) \rightarrow \neg(x \Leftrightarrow z)$

(6) $\forall x \in \Gamma\ \exists y \in \Gamma\ x \Leftrightarrow y$

Remark: In exchange algebra the selection of an exchange basis on which an exchange relation is defined is important. We do not assume that an exchange relation is preserved in a transformation between the basis. The following figure shows the relation among a bookkeeping system and its abstractions.

[Definition 5.3.9] Trivial Basis

Let Λ be an exchange basis of an exchange algebra Ψ. If $\forall x, y \in \Lambda\ \neg(x \Leftrightarrow y)$, then Ψ is called trivial and Λ is called a trivial exchange basis.

[Proposition 5.3.5]

Let Λ be an non-empty exchange basis of an exchange algebra Ψ.

$\forall x \in \Gamma\ x \Leftrightarrow \wedge x,$ where $\Gamma = \Lambda \cup \wedge\Lambda$

Proof:

We assume that $\neg(\forall x \in \Gamma\ x \Leftrightarrow \wedge x)$. Then $\exists x \in \Gamma\ \neg(x \Leftrightarrow \wedge x)$ holds. From axiom (6) of Exchange Algebra $\exists y\ y \Leftrightarrow x$ and $y \neq \wedge x$ hold.

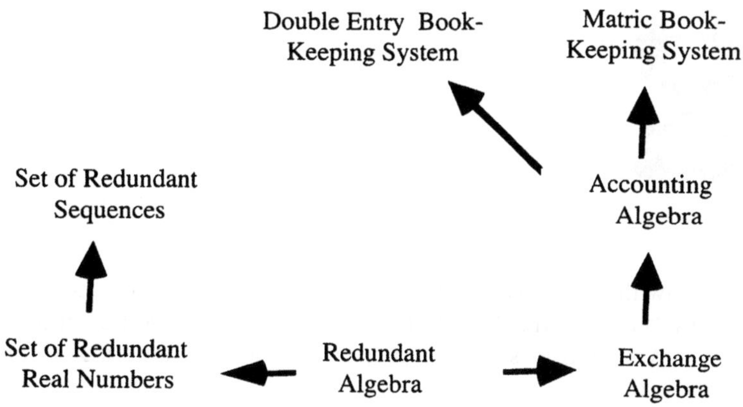

FIG. 5.2 Abstract Structure of Bookkeeping Systems

Then $\neg(x \Leftrightarrow {}^\wedge y)$ holds from axiom (4) and $\neg({}^\wedge x \Leftrightarrow {}^\wedge y)$ holds from axiom (5). However ${}^\wedge x \Leftrightarrow {}^\wedge y$ from axiom (1). This is a contradiction. Thus $x \Leftrightarrow {}^\wedge x$ holds.

Q.E.D.

[Proposition 5.3.6]

(1) $\forall x \in \Gamma\ \neg(x \Leftrightarrow x)$ (2) $\forall x, y \in \Gamma\ \neg(x \Leftrightarrow {}^\wedge y) \rightarrow x \Leftrightarrow y$

Proof:

(1) Let $x \Leftrightarrow x$. Then $\neg(x \Leftrightarrow x)$ holds from axiom (2). This is a contradiction.

(2) Let $\neg(x \Leftrightarrow {}^\wedge y)$ and $\neg(x \Leftrightarrow y)$. Then $\neg({}^\wedge y \Leftrightarrow y)$ holds from axiom (3) and (5).

This is a contradiction from Proposition 5.3.5. Q.E.D.

[Example 5.3.3] Accounting Algebra

Let Λ = {<assets>, <liability>, <equity>, <profit>, <cost>} and ^Λ = {^<assets>, ^<liability>, ^<equity>, ^<profit>, ^<cost>} and $\Gamma = \Lambda \cup {}^\wedge\Lambda$.

Let $[\Gamma]_f$ denote the commutative free semi-group generated by Γ with unit element 0. Then the operations "—" and "^" can be defined on $[\Gamma]_f$ and it is easy to show that $[\Gamma]_f$ is a redundant algebra. If the exchange relation ⇔ on Γ is defined as follows, then $[\Gamma]_f$ becomes an exchange algebra and also accounting algebra.

<assets> ⇔ <liability>, <assets> ⇔ <equity>, <assets> ⇔ <equity>, <cost> ⇔ <liability>, <cost> ⇔ <equity>, <cost> ⇔ <equity>, and if $x \Leftrightarrow y$ then $y \Leftrightarrow x$ holds. Other relations on Λ do not hold. Relations on Γ are determined uniquely from the next proposition.

[Proposition 5.3.7]

Let Λ be a basic basis of a redundant space Ψ and ⇔ be a relation on Λ which satisfy the axioms (2), (3), (5), (6). Then the relation ⇔ can be extended to the exchange relation on the extended basis Γ uniquely.

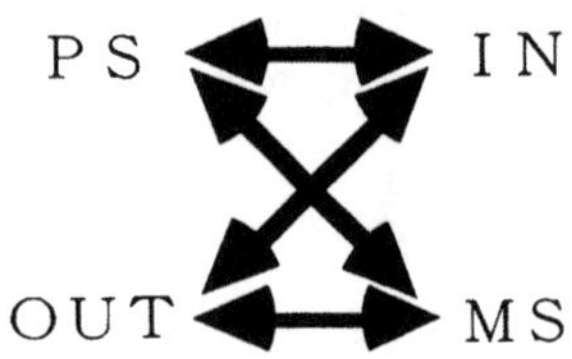

FIG. 5.3 Exchange Relation on Λ'

Proof:
At first we show that the extension is unique. Let $x \Leftrightarrow_1 y$ and $\neg(x \Leftrightarrow_2 y)$ for $x \in \Lambda, y \in {}^\wedge\Lambda$. $\neg(x \Leftrightarrow_1 {}^\wedge y)$ and $x \Leftrightarrow_2 y$ hold from the axiom (4) and Proposition 5.3.6. Where $x, {}^\wedge y \in \Lambda$, then this is a contradiction for the uniqueness of $\Leftrightarrow$ on Λ. Thus the extension is unique.

Then we define that if $x \Leftrightarrow y$ then $\neg(x \Leftrightarrow {}^\wedge y)$ and that if $\neg(x \Leftrightarrow y)$ then $x \Leftrightarrow {}^\wedge y$.

Where the exchange relation between Λ and ${}^\wedge\Lambda$ is defined and also the exchange relation on Γ is defined uniquely. Q.E.D.

[Proposition 5.3.8]
Let $\Lambda' = \{PS, IN, MS, OUT\}$.

Let $g: \Lambda \to \Lambda'$ be an onto function.

Let $\Leftrightarrow$ be a relation on Λ, which satisfy the following conditions (1) to (5). Then the relation $\Leftrightarrow$ can be extended to the exchange relation on the extended basis Γ uniquely.

(1) If $g(x) = PS$ and $g(y) = IN$ then $x \Leftrightarrow y$ and $y \Leftrightarrow x$
(2) If $g(x) = PS$ and $g(y) = MS$ then $x \Leftrightarrow y$ and $y \Leftrightarrow x$
(3) If $g(x) = OUT$ and $g(y) = IN$ then $x \Leftrightarrow y$ and $y \Leftrightarrow x$
(4) If $g(x) = OUT$ and $g(y) = MS$ then $x \Leftrightarrow y$ and $y \Leftrightarrow x$
(5) Otherwise $\neg x \Leftrightarrow y$

Proof:
Exchange Relation on Λ is defined uniquely from the definition. We assume such relation on Λ' that PS $\Leftrightarrow$ IN, IN $\Leftrightarrow$ PS, PS $\Leftrightarrow$ MS, MS $\Leftrightarrow$ PS, OUT $\Leftrightarrow$ IN, IN $\Leftrightarrow$ OUT, OUT $\Leftrightarrow$ MS and MS $\Leftrightarrow$ OUT. Otherwise $\neg(x \Leftrightarrow y)$.

Figure 5.3 shows the relation.

It is easy to show that $\Leftrightarrow$ satisfies axiom (2), (3), (5) and (6). Then $\Leftrightarrow$ can be extended to the exchange relation on the extended basis Γ' uniquely from Proposition 5.3.7.

Then the relation $\Leftrightarrow$ can be extended on Λ by the function g. Then it is easy to show that $\Leftrightarrow$ becomes the exchange relation on Γ by the same way. Q.E.D.

In the Proposition 5.3.8 PS, IN, MS and OUT mean plus stock, input, minus stock and output respectively. This gives standard interpretation of exchange algebra.

[Definition 5.3.10] Redundant Homomorphism
Let Ψ_1 and Ψ_2 are redundant algebra. Then redundant homomorphism $T: \Psi_1 \to \Psi_2$ is defined as follows.

(1) $T(\hat{}x) = \hat{}T(x)$
(2) $T(\overline{}x) = \overline{}T(x)$
(3) $T(ax) = aT(x)$
(4) $T(x + y) = T(x) + T(y)$

[Definition 5.3.11] Exchange Homomorphism
Let $(\Lambda_1, \Leftrightarrow_1)$ and $(\Lambda_2, \Leftrightarrow_2)$ are the bases of exchange algebras and its exchange relations.

Then $f: \Lambda_1 \to \Lambda_2$ is called exchange preserving map iff $\forall\ x, y \in \Lambda_1\ x \Leftrightarrow_1 y \to f(x) \Leftrightarrow_2 f(y)$.

Let T be a redundant homomorphism and also exchange preserving map then T is called exchange homomorphism.

[Definition 5.3.12] Aggregation
Let Ψ_1 and Ψ_2 are exchange algebra with accounting norm. Let Λ_1 and Λ_2 are the bases of Ψ_1 and Ψ_2 respectively.

If redundant homomorphism $T: \Psi_1 \to \Psi_2$ satisfy the following two conditions then T is called aggregation.

(1) $T: \Lambda_1 \to \Lambda_2$ is an onto mapping.
(2) $\forall x \in \Psi_1\ |x| = |T(x)|$

[Definition 5.3.13] R-L Decomposition, P-M Decomposition and Projection
Let $\Gamma = \Lambda \cup \hat{}\Lambda$ and $\Lambda \cap \hat{}\Lambda = \phi$ and Γ = debit side $\cup$ credit side and debit side $\cap$ credit side = ϕ.

Let $V_{ac} = [\Gamma]$, V_{acr} = [credit side], V_{acl} = [debit side], $V_{ac^+} = [\Lambda]$ and $V_{ac^-} = [\hat{}\Lambda]$.

(1) $x \in V_{ac}$ is decomposed uniquely as $x = x_r + x_l$, $x_r \in V_{acr}$ and $x_l \in V_{acl}$ by Proposition 5.3.3.

Then R(x) and L(x) are defined as $R(x) = x_r$ and $L(x) = x_l$. This is called R-L decomposition.

(2) $x \in V_{ac}$ is decomposed uniquely as $x = x^+ + x^-$, $x^+ \in V_{ac^+}$ and $x^- \in V_{ac^-}$.
Then P(x) and M(x) are defined as $P(x) = x^+$ and $M(x) = x^-$. This is called P-M decomposition.

(3) Let $x = \sum_{e_i \in \Gamma} a_i e_i$ then $\text{Project}[e_k](x) = a_k e_k$ is defined as projection operators.
For any $A \subseteq \Gamma$ Project[A](x) is defined as $\text{Project}[A](x) = \sum_{e \in A} \text{project}[e](x)$.

[Cor. 5.3.1]
(1) $RR = R$, $LL = L$, $PP = P$, $MM = M$
(2) $P(x+y) = P(x) + P(y)$, $P(ax) = aP(x)$, $M(x+y) = M(x) + M(y)$, $M(ax) = aM(x)$
(3) $R(x+y) = R(x) + R(y)$, $R(ax) = aR(x)$, $L(x+y) = L(x) + L(y)$, $L(ax) = aL(x)$
(4) $R(\hat{}x) = L(x)$, $L(\hat{}x) = R(x)$

[Definition 5.3.14] Norm Balance

(1) $x \in V_{ac}$ is called norm balanced iff $|R(x)| = |L(x)|$.

(2) Let $V_{nb} = \{x \mid x \in V_{ac} \text{ and } |L(x)| = |R(x)|\}$ then V_{nb} is called exchange balanced space.

[Proposition 5.3.9]

V_{nb} is a sub algebra of V_{ac}, where the following conditions hold.

(1) If $x, y \in V_{nb}$ then $x + y \in V_{nb}$.

(2) If $x \in V_{nb}$ and $a \in T$ then $ax \in V_{nb}$.

(3) If $x \in V_{nb}$ then $\hat{}x \in V_{nb}$.

(4) If $x \in V_{nb}$ then $\bar{x} \in V_{nb}$.

Proof:

(1) $|R(x) + R(y)| = |R(x)| + |R(y)|$ and $|L(x) + L(y)| = |L(x)| + |L(y)|$ hold.
From $x,y \in V_{nb}$ $|R(x)| = |L(x)|$ and $|R(y)| = |L(y)|$ hold.
Then $|R(x + y)| = |R(x) + R(y)| = |R(x)| + |R(y)| = |L(x)| + |L(y)| = |L(x) + L(y)| = |L(x + y)|$.

(2) $|R(ax)| = a\,|R(x)|$ and $|L(ax)| = a|L(x)|$ hold. Then $|R(ax)| = |L(ax)|$ holds because $|R(y)| = |L(y)|$ holds.

(3) $R(\hat{}x) = L(x)$ and $L(\hat{}x) = R(x)$ hold. Then $|R(\hat{}x)| = |L(x)| = |R(x)| = |L(\hat{}x)|$ holds.

(4) $|\bar{R}(x)| = |\bar{L}(x)|$ is shown from unique expression of x by the given bases.

Q.E.D.

The proposition means that the standard operations of a bookkeeping system are closed on V_{nb}. This is used for checking calculation errors in a bookkeeping system.

The calculations in the bookkeeping system are closed in sub algebra, norm balanced space, then we can find the error by checking the norm balanced property of the element x which is the result of any calculations on the norm balanced space.

It is possible to extend our algebra to the operator algebra of the space. It will become a functional analysis with exchange algebra. This extension is necessary for studying the problem of production and decision making. In this manuscript we omit these problems, instead we formulate the concept of an economic field.

[Multi–Dimensional and Multi–Currency Extension of Exchange Algebra]

We introduce bookkeeping system with material flow information, that is called multi–dimensional bookkeeping system on exchange algebra, by extending exchange algebra with units for materials. We also introduce multicurrency description of bookkeeping system on exchange algebra.

[Definition 5.3.15] Multi Currency Description
Let $\Gamma = \Lambda \cup {}^\wedge\Lambda$ = debit side ∪ credit side. Let Δ = {yen, dollar, euro, MOU, . . .} be a set of units of currency. Where MOU denotes the abstract unit of currency we use in this book.

Then exchange algebra $[\Gamma \times \{\alpha\}]$ for $\alpha \in \Delta$ is easily defined depending on original exchange algebra $[\Gamma]$ and called exchange algebra with currency α.

Next chapter we use exchange algebra with currency "MOU" such as <e, MOU> and ^<e, MOU>.

[Definition 5.3.16] Multi–Dimensional Description and Price Function
We introduce units for materials as follows. We use WHU (Wheat Unit) for wheat, FLU (Flour Unit) for wheat flour, BRU (Bread Unit) for bread, STU (Steel Unit) for steel and MAU (Machine Unit) for machinery. Next chapter we use multi dimensional description such as <Wheat, WHU> and ^<Wheat, WHU>.

Price function "g" is introduced for describing the price of product per unit such as g(WHU) = 4, that means the price of wheat is 4 MOU per unit. g[t] denotes the price of term t such as g[t](WHU), that means the price of wheat per unit at term t.

G is defined as a price convert function that convert multi–dimensional description to standard currency description such as G[WHU](x<Wheat, WHU>) = g(WHU)x<Wheat, MOU> = 4x<Wheat, MOU>. Unit convert function H is defined as the inverse function of G. For example H[WHU](y<Wheat, MOU>) = (y/g(WHU))<Wheat, WHU> = y/4<Wheat, WHU>.

[Definition 5.3.17]
$F(\Omega, \Gamma) = \{f \mid f: \Omega \to [\Gamma]\}$ is called economic field and $f \in F(\Omega, \Gamma)$ is called a state of the economic field.

$F(\Omega, V_{nb}) = \{f \mid f: \Omega \to V_{nb}\}$ is called norm balanced economic field.

Dynamical system on economic field is also introduced as $f(t)(\omega)$, $\omega \in \Omega$.

G and H defined above can be extended on an economic field.

[Example 5.3.4] Economic Field
Let Ω = {Agriculture, Flour–milling Industry, Bread Industry, Steel Industry, Machinery Industry, Household, Government, Bank, Central Bank, Overseas}.

Then $F(\Omega, \Gamma) = \{f \mid f: \Omega \to [\Gamma]\}$ is an economic field. Economic states of an economic system and flow quantities as transactions among agents can be described by using economic field.

For example we assume that Agriculture purchases machinery from Machinery Industry for 44 MOU. Then this transaction $f \in F(\Omega, \Gamma)$ is describe as follows.

f(Agriculture) = 44<Machinery, MOU> + 44^<Cash, MOU>
f(Machinery Industry) = 44^<Machinery, MOU> + 44<Cash, MOU>
f(ω) = 0 if $\omega \neq$ Agriculture & $\omega \neq$ Machinery Industry

5.4 Concluding Remarks

In the following chapter, a small national economy called a virtual economy is treated as a simple example of an economic field.

We analyze how the state description of an economic agent is given by exchange algebra. In addition, we analyze what kind of structure the economic field has. We give a constructive definition of macro variables of the economic system such as GDP, Investment, Savings and Consumption from a micro state description given by exchange algebra.

The dynamics of an economic field are treated in the following chapter by the analysis of a gaming simulation of a virtual economy.

Chapter 6
The National Economic Field and Algebraic Reconstruction of SNA

Aim of This Chapter

In this chapter we will give a concrete state description of an individual economic agent based on the concept of economic exchange, construct the state of a national economy from a bottom up point of view and show how the total state of a national economy consists of the state of individual economic agents. Macroeconomic variables such as Y, I, C and S make up the micro variables of economic transactions. The micro variables of economic transactions are described by a bookkeeping system for each agent.

We will construct a model of a small national economy called a virtual economy with economic sectors represented by nine autonomous agents. The micro macro structure of this small mational economy is analyzed in this chapter. However, except for the definition of a virtual economy, to understand the following chapters the formal theory of this chapter is not indispensable.

6.1 The Virtual Economy

In this section we will construct a simple agent based model of a national economy, which is called a virtual economy. A virtual economy consists of nine agents such as agriculture, a flour milling industry, a bread industry (bakery), a steel industry, a machine industry, government, household, a bank and a central bank. For describing the agent based economic state of stock and flow we have already given an algebraic abstraction of the bookkeeping system defined in the previous section, known as exchange algebra.

The economic state of each agent is described by exchange algebra. Exchange algebra is the extension of accounting vector space and by using this algebra we describe the systemic properties of economic exchange and the properties of an economic field, which gives a formal model for the SNA (System of National Accounts). We give an algebraic reconstruction of the SNA with exchange algebra and the virtual economy is illustrated as follows.

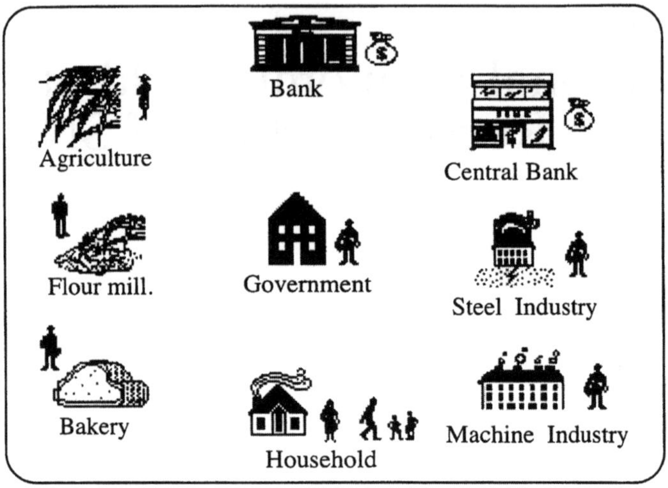

FIG. 6.1 Virtual Economy

In the virtual economy agriculture grows wheat, the flour-milling industry mills wheat flour from wheat, the bread industry (bakery) bakes bread from the flour, the steel industry forges steel and the machinery industry makes machinery from the steel. In the model we have assumed that there are no materials for a steel industry. Households purchase and consume bread. A machine is purchased by each industry as an investment and used for production. Government or households also purchase machines. The machines that are purchased by the Government or households are considered as an infrastructure and a house respectively. A machine depreciates according to the scenario and population increases. Households supply workers to each industry and the Government and households receive a wage. The Government can issue national bonds. The Central Bank issues a bank note and fixes the official bank rate. Households and industry deposit money in the bank. The bank lends money.

6.2 The Structure of Economic Field and SNA

In this chapter we will introduce the definition of a simplified national economic field.

Let Ω be a set of economic units and $[\Gamma]_{nb}$ be a norm balanced space.

Then $K(\Omega, [\Gamma]_{nb}) = \{f \mid f: \Omega \to [\Gamma]_{nb}\}$ is called a morm balanced economic field and an element f is called a state of an economic field. In the following section we usually omit "nb" for convenience.

We consider that the state description by using an economic field is common to any economic system.

[Definition 6.2.1] Algebraic Bases for the Virtual Economy
(1-a) Bases for Industry
Λ[Industry: Assets] = {<Cash, MOU>, <Deposits, MOU>,
<National Bonds, MOU>, <Steel, MOU>, <Machinery, MOU>,
<Wheat flour, MOU>, <Wheat, MOU>, <Bread, MOU>,
<Stock Investment, MOU>, <Equipment investments, MOU>}
Λ[Industry: Equity] = {<Capital stock, MOU>}
Λ[Industry: Liability]
= {<Loans payable, MOU>, <Reserve for depreciation, MOU>}
Λ[Industry: Cost] = {<Depreciation, MOU>, <Wages expenditure, MOU>,
<Interest expense, MOU>, <Taxes expense, MOU>}
Λ[Industry: Revenue] = {<Value added, MOU>, <Retained Earnings, MOU>,
<Subsidy Income, MOU>, <National Bond Interest earned, MOU>,
<Deposit Interest earned, MOU>, <Gross Profit, MOU>,
<Ordinary Profit, MOU>}

Λ[Industry] = Λ[Industry: Assets] ∪ Λ[Industry: Equity] ∪
Λ[Industry: Liability] ∪ Λ[Industry: Cost] ∪ Λ[Industry: Revenue]
^Λ[Industry] = {^e | e ∈ Λ[Industry]}
Γ[Industry] = Λ[Industry] ∪ ^Λ[Industry]

If Λ[Industry] is classified by Λ′ = {PS, IN, MS, OUT} then the exchange relation of Γ[Industry] is uniquely determined as is shown in the previous chapter.
In the following sections we assume that Λ[Industry: Assets] corresponds to PS, Λ[Industry: Equity] and Λ[Industry: Liability] correspond to MS, Λ[Industry: Cost] corresponds to OUT and Λ[Industry: Revenue] corresponds to IN.

(1-b) Bases for Industry: Multi Dimensional Extension
We now introduce a proper unit for products such as steel, machine, wheat, flour and bread as follows.

Λ[Industry: Multi Dimensional Assets]
= {<Steel, STU>, <Machinery, MAU>, <Wheat, WHU>, <Wheat flour, FLU>,
<Bread, BRU>}
Λ[Industry: Multi Dimension]
= Λ[Industry: Multi Dimension Assets] ∪ Λ[Industry: Assets] ∪
Λ[Industry: Equity] ∪ Λ[Industry: Liability] ∪ Λ[Industry: Cost] ∪
Λ[Industry: Revenue] ∪ Λ[Industry: Transfer]
^Λ[Industry: Multi Dimension] = {^e | e ∈ Λ[Industry: Multi Dimension]}
Γ[Industry: Multi Dimension]
= Λ[Industry: Multi Dimension] ∪ ^Λ[Industry: Multi Dimension]

By using a multi dimensional description we can describe a change of price caused by infration for example.

(2) Bases for Bank
Λ[Bank] = Λ[Bank: Assets] ∪ Λ[Bank: Equity] ∪ Λ[Bank: Liability] ∪
Λ[Bank: Cost] ∪ Λ[Bank: Revenue]

Λ[Bank: Assets]
= {<Loans receivable, MOU>, <Reserve deposit receivable, MOU>, <Cash, MOU>}
Λ[Bank: Equity] = {<Capital stock, MOU>}
Λ[Bank: Liability] = {<Deposits payable, MOU>, <Loans payable, MOU>}
Λ[Bank: Cost] = {<Taxes expense, MOU>, <Interest expense, MOU>}
Λ[Bank: Revenue] = {<Retained Earnings, MOU>, <Interest earned, MOU>, <Subsidy Income, MOU>, <Ordinary Profit, MOU>}

(3) Bases for the Central Bank
Λ[Central Bank] = Λ[Central Bank: Assets] ∪ Λ[Central Bank: Equity] ∪ Λ[Central Bank: Liability] ∪ Λ[Central Bank: Cost] ∪ Λ[Central Bank: Revenue]
Λ[Central Bank: Assets] = {<Gold, MOU>, <Loans receivable, MOU>, <National Bonds, MOU>, <Cash, MOU>}
Λ[Central Bank: Equity] = {<Capital stock, MOU>}
Λ[Central Bank: Liability] = {<Reserve deposit payable, MOU>, <Central Bank note payable, MOU>}
Λ[Central Bank: Cost] = {<Central Bank Payment Expense, MOU>
Λ[Central Bank: Revenue] = {<Interest earned, MOU>}

(4) Bases for Households
Λ[Household] = Λ[Household: Assets] ∪ Λ[Household: Equity] ∪ Λ[Household: Liability] ∪ Λ[Household: Cost] ∪ Λ[Household: Revenue]
Λ[Household: Assets] = {<Cash, MOU>, <Deposits, MOU>, <National Bonds, MOU>, <Machinery, MOU>}
Λ[Household: Equity] = {<Capital stock, MOU>}
Λ[Household: Liability]
= {<Loans payable, MOU>, <Reserve for depreciation, MOU>}
Λ[Household: Cost] = {<Depreciation, MOU>, <Interest expense, MOU>, <Taxes expense, MOU>, <Consumption expenditure, MOU>}
Λ[Household: Revenue] = {<Retained Earnings, MOU>, <Subsidy Income, MOU>, <Interest earned, MOU>, <Wages earned, MOU>, <Ordinary Profit, MOU>}

(5) Bases for Governments
Λ[Government] = Λ[Government: Assets] ∪ Λ[Government: Equity] ∪ Λ[Government: Liability] ∪ Λ[Government: Cost] ∪ Λ[Government: Revenue]
Λ[Government: Assets] = {<Government service, MOU>, <Cash, MOU>}
Λ[Government: Equity] = {<Capital stock, MOU>}
Λ[Government: Liability]
= {<National Bonds payable, MOU>, <Loans payable, MOU>, <Reserve for depreciation, MOU>}
Λ[Government: Cost] = {<Depreciation, MOU>, <Wages expenditure, MOU>, <Interest expense, MOU>, <Consumption expenditure, MOU>},

<Subsidy Expense, MOU>}
Λ[Government: Revenue] = {<Value added, MOU>,
<Retained Earnings, MOU>, <Interest earned, MOU>,
<Taxes revenue, MOU>, <Central Bank Payment Income, MOU>,
<Ordinary Profit, MOU>}

(6) Bases for the Oversea Sector
Within the virtual economy framework we can add export and import transactions as needed.

(7) Bases for the Virtual Economy
Λ[Virtual Economy] = Λ[Industry] ∪ Λ[Bank] ∪ Λ[Central Bank] ∪
Λ[Household] ∪ Λ[Government] ∪ Λ[Oversea Sector]
Λ[Virtual Economy: Multi Dimension]
= Λ[Industry: Multi Dimension] ∪ Λ[Virtual Economy]
Γ[Virtual Economy] = Λ[Virtual Economy] ∪ ^Λ[Virtual Economy]
Γ[Virtual Economy: Multi Dimension]
= Λ[Virtual Economy: Multi Dimension]∪^Λ[Virtual Economy:
Multi Dimension]

(8) Exchange Algebra and Economic Field Bases for the Virtual Economy
Exchange algebra constructed from Γ[Virtual Economy] and Γ[Virtual Economy: Multi Dimension] are expressed by [Γ[Virtual Economy]] and [Γ[Virtual Economy: Multi Dimension]]. Economic field on bases for virtual economy is defined as follows.

F(Ω, Γ[Virtual Economy]) = {f | f: Ω → [Γ[Virtual Economy]]}

[Description of Transactions]
A transaction is described by an accounting vector.

For example let x = 20^<Wheat, MOU> + 20<Cash, MOU>.

This transaction is described by the standard bookkeeping system as follows.

Credit Side	**Debit Side**
Cash 20	Wheat 20

Table 6.2.1 describes the following accounting vector.

TABLE 6.2.1 An Example of a Balance Sheet

Credit Side			Debit Side		
Cash	260.16	MOU	Capital Stock	100	MOU
Machinery	99	MOU	Retained Earnings	357.76	MOU
Deposits	137	MOU	Loans Payable	0	MOU
National Bonds	10	MOU	Reserve for Depreciation	48.4	MOU
Wheat(Stock)	0	MOU			
Total	506.16	MOU		506.16	MOU

f[Initial Stock](Flour-milling Industry)
= 260.16<Cash, MOU> + 100.00<Capital stock, MOU>
+ 99.00<Machinery, MOU> + 357.76<Retained Earnings, MOU>
+ 137.00<Deposits, MOU> + 0.00<Loans payable, MOU>
+ 10.00<National Bonds, MOU> + 48.40<Reserve for depreciation, MOU>
+ 0.00<Wheat, MOU> ∈ [Γ[Virtual Economy]] (6.2.1)

Project[Credit](f[Initial Stock] (Flour-milling Industry))
= 260.16<Cash, MOU> + 99.00<Machinery, MOU> + 137.00<Deposits, MOU>
+ 10.00<National Bonds, MOU> + 0.00<Wheat, MOU>
|Project[Credit](f[Initial Stock](Flour-milling Industry))| = 506.16
Project[Debit](f[Initial Stock](Flour-milling Industry))
= 100.00<Capital stock, MOU> + 357.76<Retained Earnings, MOU>
+ 0.00<Loans payable, MOU> + 48.40<Reserve for depreciation, MOU>
|Project[Debit](f[Initial Stock](Flour-milling Industry))| = 506.16

f_m[Machinery Purchase] = 4<Machinery, MAU> + 44^<Cash, MOU> (6.2.2)

f[Machinery Purchase] = G(f_m[Machinery Purchase])
= 44<Machinery, MOU> + 44^<Cash, MOU> (6.2.2)′

f_m[Machinery Scrap]
= 1^<Machinery, MAU> + 10^<Reserve for depreciation, MOU> (6.2.3)

f[Machinery Scrap] = G(f_m[Machinery Scrap])
= 10^<Machinery, MOU>
+ 10^<Reserve for depreciation, MOU> (6.2.3)′

f_m[Wheat Production] = 1320<Wheat, WHU> + 330<Value added, MOU>
f[Wheat Production] = G(f_m[Wheat Production])
= 330<Wheat, MOU> + 330<Value added, MOU> (6.2.4)′

Where "G" is the price convert function defined on Chapter 5.

6.3 Classification of Transactions

6.3.1 Introduction

Transactions are classified as follows.

(1) Price: Decision Making
(2) Material Purchase, Production, Selling: Decision Making
(3) Equipment investments: Decision Making
(4) Employment: Decision Making
(5) Financing: Decision Making
(6) Financial Parameters: Decision Making
(7) Government: Policy Making

The following decisions are determined as conventional accompanying decision making.

(1) Government Production Transaction and Consumption expenditure,
(2) Household Consumption expenditure
(3) Depreciation
(4) Value added Adjustment for price fluctuations of product stock
(5) Interest and Tax as an income transfer
(6) Bank Note Issue and Offset Transactions
(7) Transfer transactions

Let Industry = {Agriculture, Flour-milling industry, Bread Industry, Steel Industry, Machinery Industry}.
We treat the products of the industry in virtual economy as the following set.
Products = {<Steel, MOU>, <Machinery, MOU>, <Wheat flour, MOU>, <Wheat, MOU>, <Bread, MOU>}. We introduce the set of name of products such as Name of Products = {Steel, Machinery, Wheat flour, Wheat, Bread}.
We assume that each industry makes single product from single material in this chapter.
Then the following functions are defined.
Name: Products → Name of Products is introduced such as Name(<Steel, MOU>) = Steel.
WhatProduct: Industry → Name of Products is introduced such as WhatProduct(Agriculture) = Wheat.
Let e ∈ Name of Products then,

WhatMaterial: Name of Products → Name of Products is introduced such as WhatMaterial(Machinery) = Steel. If a product dose not need any material then value of WhatMaterial becomes null set such as WhatMaterial(Steel) = ϕ.

[Initial Stock of Industry]
Let ω ∈ Industry, ep(ω) = WhatProduct(ω), em(ω) = WhatMaterial(ep(ω)).
f[Initial Stock(ω, t) ∈ F(Ω, Γ[Virtual Economy])
= InitStock[Cash](ω, t)<Cash, MOU>
+ InitStock[ep(ω)](ω, t)<ep(ω), MOU>
+ InitStock[em(ω)](ω, t)<em(ω), MOU>
+ InitStock[Machinery](ω, t)<Machinery, MOU>
+ InitStock[Deposits](ω, t)<Deposits, MOU>
+ InitStock[National Bonds](ω, t)<National Bonds, MOU>
+ InitStock[Capital stock](t)<Capital stock, MOU>
+ InitStock[Loans payable](ω, t)<Loans payable, MOU>
+ InitStock[Retained Earnings](ω, t)<Retained Earnings, MOU>
+ InitStock[Reserve for depreciation](ω, t)<Reserve for depreciation, MOU>

Where InitStock[e](ω, t)∈T and |Project[Credit](f[Initial Stock(ω, t))| = |Project[Debit](f[Initial Stock](ω, t)))| holds.

[Households Initial Stock]
f[Initial Stock(Household, t)
= InitStock[Cash](Household, t)<Cash, MOU>
+ InitStock[Machinery](Household, t)<Machinery, MOU>

+ InitStock[Deposits](Household, t)<Deposits, MOU>
+ InitStock[National Bonds](Household, t)<National Bonds, MOU>
+ InitStock[Capital stock](t)<Capital stock, MOU>
+ InitStock[Loans payable](Household, t)<Loans payable, MOU>
+ InitStock[Retained Earnings](Household, t)<Retained Earnings, MOU>
+ InitStock[Reserve for depreciation](Household, t)<Reserve for depreciation, MOU>

[Government Initial Stock]
f[Initial Stock(Government, t)
= InitStock[Cash](Government, t)<Cash, MOU>
+ InitStock[Machinery](Government, t)<Machinery, MOU>
+ InitStock[Capital stock](t)<Capital stock, MOU>
+ InitStock[Retained Earnings](Government, t)<Retained Earnings, MOU>
+ InitStock[Reserve for depreciation](Government, t)
<Reserve for depreciation, MOU>
+ InitStock[National Bonds payable: Agriculture](Government, t)
<National Bonds payable, MOU>
+ InitStock[National Bonds payable: Flour-milling Industry]
(Government, t)<National Bonds payable, MOU>
+ InitStock[National Bonds payable: Bread Industry]
(Government, t)<National Bonds payable, MOU>
+ InitStock[National Bonds payable: Steel Industry](Government, t)
<National Bonds payable, MOU>
+ InitStock[National Bonds payable: Machinery Industry]
(Government, t)<National Bonds payable, MOU>
+ InitStock[National Bonds payable: Household](Government, t)
<National Bonds payable, MOU>
+ InitStock[National Bonds payable: Central Bank](Government, t)
<National Bonds payable, MOU>

[Bank Initial Stock]
f[Initial Stock(Bank, t)
= InitStock[Cash](Bank, t)<Cash, MOU>
+ InitStock[Loans receivable: Agriculture](Bank, t)<Loans receivable, MOU>
+ InitStock[Loans receivable: Flour-milling Industry](Bank, t)
<Loans receivable, MOU>
+ InitStock[Loans receivable: Bread Industry](Bank, t)
<Loans receivable, MOU>
+ InitStock[Loans receivable: Steel Industry](Bank, t)
<Loans receivable, MOU>
+ InitStock[Loans receivable: Machinery Industry](Bank, t)
<Loans receivable, MOU>
+ InitStock[Loans receivable: Household](Bank, t)<Loans receivable, MOU>
+ InitStock[Central Bank Reserve Fund Receivable](Bank, t)<Reserve deposit receivable, MOU>

+ InitStock[Capital stock](t)<Capital stock, MOU>
+ InitStock[Retained Earnings](Bank, t)<Retained Earnings, MOU>
+ InitStock[Deposits payable: Agriculture](Bank, t)<Deposits payable, MOU>
+ InitStock[Deposits payable: Flour-milling Industry](Bank, t) <Deposits payable, MOU>
+ InitStock[Deposits payable: Bread Industry](Bank, t) <Deposits payable, MOU>
+ InitStock[Deposits payable: Steel Industry](Bank, t) <Deposits payable, MOU>
+ InitStock[Deposits payable: Machinery Industry](Bank, t) <Deposits payable, MOU>
+ InitStock[Deposits payable: Household](Bank, t)<Deposits payable, MOU>
+ InitStock[Central BankLoans payable](Bank, t)<Loans payable, MOU>

[Central Bank Initial Stock]
f[Initial Stock(Central Bank, t)
= InitStock[Gold](Central Bank, t)<Gold, MOU>
+ InitStock[Central BankLoans receivable](Central Bank, t) <Loans receivable, MOU>
+ InitStock[National Bonds](Central Bank, t)<National Bonds, MOU>
+ InitStock[Capital stock](t)<Capital stock, MOU>
+ InitStock[Central Bank Reserve Fund Payable](Central Bank, t) <Reserve deposit payable, MOU>
+ InitStock[Central Bank note payable](Central Bank, t) <Central Bank note payable, MOU>

6.3.2 Main Decision Making

(6.3.2.1) [Price[e](t): Decision Making]

Price[e](t), $e \in$ Titles of Products are determined by the market interactions among agents. Price[Wheat](t), Price[Wheat flour](t), Price[Bread](t), Price[Steel](t), Price[Machinery](t) are determined by the market interaction.

(6.3.2.2) [Material Purchase, Production, Selling: Decision Making]

Agriculture produces wheat without any materials.
Product[Wheat](t)<Wheat, WHU> denotes the production of wheat by agriculture.
Sell[Wheat](t)<Wheat, WHU> denotes the selling of wheat to the flour-milling industry.
Product[Wheat flour](t)<Wheat flour, FLU> denotes the production of wheat flour from wheat by the flour-milling industry.
Sell[Wheat flour](t)<Wheat flour, FLU> denotes the selling of wheat flour to the bread industry.

Product[Bread](t)<Bread, BRU> denotes the production of bread from wheat flour by the bread industry.
Sell[Bread](t)<Bread, BRU> denotes the selling of bread to households.
Product[Steel](t)<Steel, STU> denotes the production of steel by the steel industry.
The steel industry also produces steel without any material.
Sell[Steel](t)<Steel, STU> denotes the selling of steel to the machinery industry.
Product[Machinery](t)<Machinery, MAU> denotes the production of machinery by the machinery industry.
Sell[Machinery](t)<Machinery, MAU> denotes the selling of machinery to agriculture, flour-milling industry, bread industry, steel industry, machinery industry, households and the government.
Export and import transactions are introduced as needed.

<Production>
f[Production: ep(ω)](ω) = Use[em(ω)](t)^<em(ω), MOU>
+ ProPrice[ep(ω)](t)<ep(ω), MOU> + ValueAdd[ω](t)<Value added, MOU>,
ProPrice[ep(ω)](t) = Use[em(ω)](t) + ValueAdd[ω](t)
Where ep(ω) = WhatProduct(ω), em(ω) = WhatMaterial(ep(ω)),
ProPrice[ep(ω)](t) = Product[ep(ω)](t) × Price[ep(ω)](t).
Use[em(ω)](t) = 0 when an industry dose not use material.

[Example of Production by Agriculture]
f[Production: Wheat](Agriculture) = ProPrice[Wheat](t)<Wheat, MOU>
+ ProPrice[Wheat](t)<Value added, MOU>
f[Production: Wheat](ω) = 0 if ω ≠ Agriculture
Where ProPrice[Wheat](t) = Product[Wheat](t) Ω Price[Wheat](t)

<Sell & Buy>
Selling from ω to α and buying from ω are defined for e = WhatProduct(ω) as follows.
f[Sell: e: α](ω) = SellPrice[e](t)^<e, MOU> + SellPrice[e](t)<Cash, MOU>
f[Buy: e: ω](α) = SellPrice[e](t)<e, MOU> + SellPrice[e](t)^<Cash, MOU>
SellPrice[e](t) = Sell[e](t) × Price[e](t)

<Domestic Intermediate Use>
(1) Wheat: Sell & Buy
f[Sell: Wheat: Flour-milling Industry](Agriculture)
= SellPrice[Wheat](t)^<Wheat, MOU>
+ SellPrice[Wheat](t)<Cash, MOU>
f[Buy: Wheat: Agriculture](Flour-milling Industry)
= SellPrice[Wheat](t)<Wheat, MOU>
+ SellPrice[Wheat](t)^<Cash, MOU>
SellPrice[Wheat](t) = Sell[Wheat](t) × Price[Wheat](t)

(2) Wheat flour: Sell & Buy
f[Sell: Wheat flour: Bread Industry](Flour-milling Industry)
= SellPrice[Wheat flour](t)^<Wheat flour, MOU>
+ SellPrice[Wheat flour](t)<Cash, MOU>,
f[Buy: Wheat flour: Flour-milling Industry](Bread Industry)
= SellPrice[Wheat flour](t)<Wheat flour, MOU>
+ SellPrice[Wheat flour](t)^<Cash, MOU>,
SellPrice[Wheat flour](t) = Sell[Wheat flour](t) × Price[Wheat flour](t)

(3) Steel: Sell & Buy
f[Sell: Steel: Machinery Industry](Steel Industry)
= SellPrice[Steel](t)^<Steel, MOU>
+ SellPrice[Steel](t)<Cash, MOU>
f[Buy: Steel: Steel Industry](Machinery Industry)
= SellPrice[Steel](t)<Steel, MOU>
+ SellPrice[Steel](t)^<Cash, MOU>
SellPrice[Steel](t) = Sell[Steel](t) × Price[Steel](t)

<Consumption Expenditure>
(4) Bread: Domestic Sell & Buy
f[Sell: Bread: Household](Bread Industry)
= SellPrice[Bread](t)^<Bread, MOU>
+ SellPrice[Bread](t)<Cash, MOU>
f[Buy: Bread: Bread Industry](Household)
= SellPrice[Bread](t)<Bread, MOU>
+ SellPrice[Bread](t)^<Cash, MOU>
SellPrice[Bread](t) = Sell[Bread](t) × Price[Bread](t)

<Export Transactions>
Let e ∈ Titles of Products, ep(ω) = WhatProduct(ω), ω ∈ Industry.
f[Export: ep(ω)](ω)
= Export[ep(ω)](ω, t)^<ep(ω), MOU> + Export[ep(ω)](ω, t)<Cash, MOU>
f[Export: ep(ω)](Oversea)
= Export[ep(ω)](ω, t)<ep(ω), MOU> + Export[ep(ω)](ω, t)^<Cash, MOU>

<Import Transactions>
Let e ∈ Titles of Products, ω ∈ Industry ∪ {Government, Household}.
f[Import: e](ω) = Import[e](ω, t)<e, MOU> + Import[e](ω, t)^<Cash, MOU>
f[Import: e](Oversea)
= Import[e](ω, t)^<e, MOU> + Import[e](ω, t)<Cash, MOU>

<Stock Investments>
Stock investments are described by transfer transactions.

Materials and products are transferred as stock investments at the end of each term and transferred again as materials and products at the start of the next term. This is a convention for accounting calculation.

(6.3.2.3) [Equipment Investments: Decision Making of Buying Machinery]

Agriculture: Invest[Agriculture](t)<Machinery, MAU>
Flour-milling Industry: Invest[Flour-milling Industry](t)<Machinery, MAU>
Bread Industry: Invest[Bread Industry](t)<Machinery, MAU>
Steel Industry: Invest[Steel Industry](t)<Machinery, MAU>
Machinery Industry: Invest[Machinery Industry](t)<Machinery, MAU>
Household: Invest[Household](t)<Machinery, MAU>
Government: Invest[Government](t)<Machinery, MAU>
Where, Sell[Machinery](t) = Invest[Agriculture](t)
+ Invest[Flour-milling Industry](t) + Invest[Bread Industry](t)
+ Invest[Steel Industry](t) + Invest[Machinery Industry](t)
+ Invest[Household](t) + Invest[Government](t).

<Domestic Equipment Investments>
Let ω ∈ Industry ∪ {Household, Government}
f[Sell: Machinery: ω](Machinery Industry)
= InvPrice[ω](t)<Cash, MOU> + InvPrice[ω](t)^<Machinery, MOU>, where ω is a buyer of Machinery.
f[Domestic Equipment investments: Machinery: ω](ω)
= InvPrice[ω](t)^<Cash, MOU> + InvPrice[ω](t)<Machinery, MOU>
InvPrice[ω](t) = Invest[ω](t) × Price[Machinery](t)

(6.3.2.4) [Employment: Decision Making for Wages and Numbers in Employment]

Let ω ∈ Industry ∪ {Government}.
Employed[Household](t) denotes the number of total employed and Wages[Household](t) denotes the total of household wages.

Employed[Household](t)
= Employ[Agriculture](t) + Employ[Flour-milling Industry](t)
+ Employ[Bread Industry](t) + Employ[Steel Industry](t)
+ Employ[Machinery Industry](t) + Employ[Government](t)

Wages[Household](t)
= Wages[Agriculture](t) + Wages[Flour-milling Industry](t)
+ Wages[Bread Industry](t) + Wages[Steel Industry](t)
+ Wages[Machinery Industry](t) + Wages[Government](t)

Where we assume that there is no employment in the bank and Central Bank.

<Employment>
Let ω ∈ Industry ∪ {Government}.
f[Employment: ω](ω)
= Wages[ω](t)<Wages expenditure, MOU> + Wages[ω](t)^<Cash, MOU>

f[Employment: ω](Household)
= Wages[ω](t)<Wages earned, MOU> + Wages[ω](t)<Cash, MOU>

(6.3.2.5) [Financing: Decision Making]

(1) Industry and Households: Deposits and Withdrawals
Saving[ω](t)<Deposits, MOU> denote the deposit to bank by ω, where ω ∈ Industry ∪ {Household}.
Saving[Bank](t) denotes the total amount of deposits from industry and household as follows.
Saving[Bank](t) = Saving[Agriculture](t) + Saving[Flour-milling Industry](t)
+ Saving[Bread Industry](t) + Saving[Steel Industry](t)
+ Saving[Machinery Industry](t) + Saving[Household](t).
Withdraw[ω](t)^<Deposits, MOU> denotes the withdrawal from the bank by ω, where ω ∈ Industry ∪ {Household}.
Withdraw[Bank](t) = Withdraw[Agriculture](t)
+ Withdraw[Flour-milling Industry](t) + Withdraw[Bread Industry](t)
+ Withdraw[Steel Industry](t) + Withdraw[Machinery Industry](t)
+ Withdraw[Household](t)

<Deposits and Withdrawal>
(a) ω ∈ Industry ∪ {Household}: Deposit Transactions
f[Deposit: ω](ω)
= Saving[ω](t)^<Cash, MOU> + Saving[ω](t)<Deposits, MOU>
f[Deposit: ω](Bank)
= Saving[ω](t)<Cash, MOU> + Saving[ω](t)<Deposits payable, MOU>
(b) ω ∈ Industry ∪ {Household}: Withdrawal Transactions
f[Withdraw Deposit: ω](ω)
= Withdraw[ω](t)<Cash, MOU> + Withdraw[ω](t)^<Deposits, MOU>
f[Withdraw Deposit: ω](Bank)
= Withdraw[ω](t)^<Cash, MOU> + Withdraw[ω](t)^<Deposits payable, MOU>

(2) Industry and Household: Loans and Repayments
Loan[ω](t)<Loans payable, MOU> denotes a loan from the bank, where ω ∈ Industry ∪ {Household}. Loan[Bank](t) denotes the total amount of the loan as follows.

Loan[Bank](t)
= Loan[Agriculture](t) + Loan[Flour-milling Industry](t)
+ Loan[Bread Industry](t) + Loan[Steel Industry](t)
+ Loan[Machinery Industry](t) + Loan[Household](t)

Repay[ω](t)^<Loans payable, MOU> denotes the repayment of loan.
Repayed[Bank](t)
= Repay[Agriculture](t) + Repay[Flour-milling Industry](t)
+ Repay[Bread Industry](t) + Repay[Steel Industry](t)
+ Repay[Machinery Industry](t) + Repay[Household](t)

<Loans and Repayments>
(a) ω ∈ Industry ∪ {Household}: Loan Transactions
f[Loan: ω](Bank)
= Loan[ω](t)<Loans receivable, MOU> + Loan[ω](t)^<Cash, MOU>
f[Loan: ω](ω)
= Loan[ω](t)<Loans payable, MOU> + Loan[ω](t)<Cash, MOU>
(b) ω ∈ Industry ∪ {Household}: Repayment Transactions
f[Repay: ω](Bank)
= Repay[ω](t)^<Loans receivable, MOU> + Repay[ω](t)<Cash, MOU>
f[Repay: ω](ω)
= Repay[ω](t)^<Loans payable, MOU> + Repay[ω](t)^<Cash, MOU>

(3) National Bonds: Issue & Redemption
The Government issues National Bonds. Industry and households can purchase these National Bonds.
NatBond[ω](t)<National Bonds, MOU> denotes the purchase of National Bonds, where ω ∈ Industry ∪ {Household}.
NatBond_Issue_PS[Government](t) = NatBond[Agriculture](t)
+ NatBond[Flour-milling Industry](t) + NatBond[Bread Industry](t)
+ NatBond[Steel Industry](t) + NatBond[Machinery Industry](t)
+ NatBond[Household](t)

RedeemNB[ω](t)^<National Bonds, MOU> denotes the redemption of National Bonds.
RedeemNB_PS[Government](t) = RedeemNB[Agriculture](t)
+ RedeemNB[Flour-milling Industry](t) + RedeemNB[Bread Industry](t)
+ RedeemNB[Steel Industry](t) + RedeemNB[Machinery Industry](t)
+ RedeemNB[Household](t)

<National Bonds: Purchase & Redemption>
(a) National Bonds Purchase
Let ω ∈ Industry ∪ {Household}.
f[National Bond Purchase: ω](ω)
= NatBond[ω](t)<National Bonds, MOU>
+ NatBond[ω](t)^<Cash, MOU>
f[National Bond Purchase: ω](Government)
= NatBond[ω](t)<National Bonds payable, MOU>
+ NatBond[ω](t)<Cash, MOU>
(b) National Bonds: Redemption
Let ω ∈ Industry ∪ {Household}.
f[National Bond Redeem: ω](ω)
= RedeemNB[ω]^<National Bonds, MOU>
+ RedeemNB[ω]<Cash, MOU>
f[National Bond Redeem: ω](Government)
= RedeemNB[ω]^<National Bonds payable, MOU>
+ RedeemNB[ω]^<Cash, MOU>

(4) National Bonds: Central Bank Underwritings & Redemption

The Central Bank may underwrite National Bonds as follows with no interest.

(a) National Bond: Central Bank Underwriting

f[National Bond Central Bank Underwriting](Central Bank)
= NatBondCB(t)<National Bonds, MOU>
+ NatBondCB(t)^<Cash, MOU>
f[National Bond Central Bank Underwriting](Government)
= NatBondCB(t)<National Bonds payable, MOU>
+ NatBondCB(t)<Cash, MOU>

(b) National Bonds: Redemption

f[National Bond Central Bank Redeem](Central Bank)
= RedeemNBCB(t)^<National Bonds, MOU>
+ RedeemNBCB(t)<Cash, MOU>
f[National Bond Central Bank Redeem](Government)
= RedeemNBCB(t)^<National Bonds payable, MOU>
+ RedeemNBCB(t)^<Cash, MOU>

(5) Loans from the Central Bank to the Bank

(a) Loans from the Central Bank

f[Central Bank Loan](Central Bank)
= CBLoan(t)<Loans receivable, MOU> + CBLoan(t)^<Cash, MOU>
f[Central Bank Loan](Bank)
= CBLoan(t)<Cash, MOU> + CBLoan(t)<Loans payable, MOU>

(b) Repayments from the Central Bank

f[Central Bank Repay](Central Bank)
= CBRepay(t)^<Loans receivable, MOU> + CBRepay(t)<Cash, MOU>
f[Central Bank Repay](Bank)
= CBRepay(t)^<Cash, MOU> + CBRepay(t)^<Loans payable, MOU>

(6) Central Bank Reserve Fund Deposit and Withdrawals

(a) Reserve Fund Deposits

f[Reserve Fund Deposit](Central Bank)
= CBReserveFund(t)<Reserve deposit payable, MOU>
+ CBReserveFund(t)<Cash, MOU>
f[Reserve Fund Deposit](Bank)
= CBReserveFund(t)^<Cash, MOU>
+ CBReserveFund(t)<Reserve deposit receivable, MOU>

(b) Reserve Fund Withdrawals from the Central Bank

f[Reserve Fund Withdraw](Central Bank)
= CBWithdraw(t)^<Reserve deposit payable, MOU>
+ CBWithdraw(t)^<Cash, MOU>
f[Reserve Fund Withdraw](Bank)
= CBWithdraw(t)<Cash, MOU>
+ CBWithdraw(t)^<Reserve deposit receivable, MOU>

(6.3.2.6) [Financial Parameters: Decision Making]

The Bank decides the deposit interest rate: Interest_Deposit(t) and loan interest: Interest_Loan(t). The Central Bank determines the official bank rate: Official_Bank_Rate(t).

(6.3.2.7) [Government: Policy Making]

The Government determines the corporation tax, income tax and national bond interest rate. CorpTaxRate(t), IncomeTaxRate(t) and NatBondRate(t) denote the corporation tax rate, income tax rate and national bond interest respectively.

Subsidy[ω](t) denotes the subsidy by government to ω, where ω ∈ Industry ∪ {Household, Bank}.

<Government: Subsidy>
f[Government Subsidy: ω](ω)
= Subsidy[ω](t)<Subsidy Income, MOU> + Subsidy[ω](t)<Cash, MOU>
f[Government Subsidy: ω](Government)
= Subsidy[ω](t)<Subsidy Expense, MOU> + Subsidy[ω](t)^<Cash, MOU>

6.3.3 Accompanying Decision Making

(6.3.3.1) [Government Production and Consumption]

(a) Government Service Production
The government produces and consumes the service. Production of government services is also government consumption expenditure. Government consumption expenditure is defined as a sum of wage. This is a convention.
f[Government Service Production](Government)
= Wages[Government](t)<Government Service, MOU>
+ Wages[Government](t)<Value added, MOU>
(b) Government: Service Expenditure
f[Government Service Expenditure](Government)
= Wages[Government](t)^<Government Service, MOU>
+ Wages[Government](t)<Consumption expenditure, MOU>

(6.3.3.2) [Household Consumption Expenditure]

f[Household_Consumption expenditure](Household)
= HouseDomCons(t)^<Bread, MOU>
+ HouseDomCons(t)<Consumption expenditure, MOU>
+ Import[Bread](Household, t)^<Bread, MOU>
+ Import[Bread](Household, t)<Consumption expenditure, MOU>
Where, HouseDomCons(t) = Sell[BRU](t) × Price[BRU](t)

(6.3.3.3) [Depreciation and Scrapping of Equipment]

Let ω ∈ Industry ∪ {Government, Household}.
f[Depreciation: ω](ω) = Dep[ω](t)<Depreciation, MOU>
+ Dep[ω](t)<Reserve for depreciation, MOU>

Where Dep[ω](t) is given by the stock of machinery and the depreciation rate.

We assume the depreciation rate is 20% for equipment investment and 10% for housing and public investment in our virtual economy model. We also assume that machinery is scrapped after 5 years and house is demolished after 10 years.

When machinery is scrapped then the machine is offset by a reserve for depreciation as follows.

f[Scrap of Equipment: ω](ω) = Scrap[ω](t)^<Machinery, MOU>
+ Scrap[ω](t)^<Reserve for depreciation, MOU>

Let g[t – 6](MAU) = Price[MAU](t – 6) then Scrap[ω](t) = Price[MAU](t – 6) × ScrapMach[ω](t). Where ScrapMach[ω] = Invest[ω](t – 6).

(6.3.3.4) [The Adjustment of Value Added Depending on the Change of Product Stock Price]

Adjustment of value added is required if the price of product stock is changed. The adjustment is given as follows.

f[Value added_Product Stock AdjustmentUp: Wheat](Agriculture)
= ValAdjustUp[Agriculture](t)<Wheat, MOU>
+ ValAdjustUp[Agriculture](t)<Value added, MOU>

Where the price of wheat of the previous term is denoted by g[t – 1](WHU).
g[t](WHU) denotes the price of this term.
ProductStock[Agriculture](t)<Wheat, WHU> denotes the product stock that is measured by the proper unit.
Then ValAdjustUp[Agriculture](t)
= Max(0, g[t](WHU) – g[t – 1](WHU)) × ProductStock[Agriculture](t)
when the price becomes high.

f[Value added_Product Stock AdjustmentDown: Wheat]
= ValAdjustDown[Agriculture](t)^<Wheat, WHV>
+ ValAdjustDown[Agriculture](t)^<Value added, MOU>

Where ProductStock[Agriculture](t)<Wheat, WHU> denotes the product stock that is measured by the proper unit.
Then ValAdjustDown[Agriculture](t) = Max(0, g[t – 1](WHU)) – g[t](WHU)) × ProductStock[Agriculture](t) when the price becomes low.

f[Value added_Product Stock Adjustment](Agriculture)
= ValadjustUp[Agriculture](t)<Wheat, MOU>
+ ValAdjustDown[Agriculture](t)^<Wheat, MOU>
+ ValadjustUp[Agriculture](t)<Value added, MOU>
+ ValAdjustDown[Agriculture](t)^<Value added, MOU>

Let ω ∈ Industry and ep(ω) = WhatProduct(ω).

f[Value added_Product Stock Adjustment](ω)
= ValadjustUp[ω](t)<ep(ω), MOU>
+ ValAdjustDown[ω](t)^<ep(ω), MOU>

+ ValadjustUp[ω](t)<Value added, MOU>
+ ValAdjustDown[ω](t)^<Value added, MOU>
ValAdjustUp[ω](t)
= Max(0, g[t](ep(ω)) – g[t – 1](ep(ω))) × ProductStock[ω](t)
ValAdjustDown[ω](t)
= Max(0, g[t – 1](ep(ω))) – g[t](ep(ω))) × ProductStock[ω](t)

A Value added Adjustment Transaction is required when the price of product stock fluctuates. We do not need to make any adjustment in the case of material stock because the effect of the price change is included in the cost of material.

(6.3.3.5) [Income Transfer by the Payment of Interest and Tax]

(1) Interest Earned for Deposit
Let ω ∈ Industry ∪ {Household}.
f[Deposit Interest: ω](ω)
= SaveInt[ω](t)<Cash, MOU> + SaveInt[ω](t)<Interest earned, MOU>
f[Deposit Interest: ω](Bank)
= SaveInt[ω](t)^<Cash, MOU> + SaveInt[ω](t)<Interest expense, MOU>
Where SaveInt[ω](t)
= Interest_Deposit(t) × (InitStock[Deposits](ω, t)
+ Saving[ω](t) – Withdraw[ω](t))
InitStock[Deposits](ω, t) denotes the initial stock of deposit at term t.

(2) Interest Expenses for Bank Loans
Let ω ∈ Industry ∪ {Household}.
f[Loan Interest: ω](Bank)
= LoanInt[ω](t)<Cash, MOU> + LoanInt[ω](t)<Interest earned, MOU>
f[Loan Interest: ω](ω)
= LoanInt[ω](t)^<Cash, MOU> + LoanInt[ω](t)<Interest expense, MOU>
Where LoanInt[ω](t)
= Interest_Loan(t) × (InitStock[Loans payable](ω, t)
+ Loan[ω](t) – Repay[ω](t)).
InitStock[Loans payable](ω, t) denotes the initial stock of Loan payable at term t.

(3) Interest Transactions for National Bonds
Let ω ∈ {Household} ∪ Industry
f[National Bond Interest: ω](ω)
= NBondInt[ω](t)<Cash, MOU> + NBondInt[ω](t)<Interest earned, MOU>
f[National Bond Interest: ω](Government)
= NBondInt[ω](t)^<Cash, MOU> + NBondInt[ω](t)<Interest expense, MOU>
NBondInt[ω](t)
= NatBondRate(t) × (InitSTNB[ω](t) + NatBond[ω](t) –RedeemNB[ω](t))
Where NatBondRate(t) denotes the interest rate of National Bonds and InitSTNB[ω](t) denotes the initial stock of National Bond at term t.

NBondInt[Government](t)
= |∑{Project[<Interest expense, MOU>](f[National Bond Interest Transaction: ω](Government)) | ω ∈ {Household} ∪ Industry}|

(4) Interest Transactions for the Official Bank Rate
f[Central Bank Loan Interest](Central Bank)
= CBLoanInt(t)<Interest earned, MOU> + CBLoanInt(t)<Cash, MOU>
f[Central Bank Loan Interest](Bank)
= CBLoanInt(t)^<Cash, MOU> + CBLoanInt(t)<Interest expense, MOU>
CBLoanInt(t) = Official_Bank_Rate(t) × (InitStock[Central BankLoans payable](Bank, t) + CBLoan(t) − CBRepay(t))
Where InitStock[Central BankLoans payable](Bank, t) denotes the initial stock of Central BankLoans payable at term t.

(5) Central Bank Payment Transactions
f[Central Bank Payment](Central Bank)
= GovReceipt(t)<Central Bank Payment Expense, MOU>
+ GovReceipt(t)^<Cash, MOU>
f[Central Bank Payment](Government) = GovReceipt(t)<Cash, MOU>
+ GovReceipt(t)<Central Bank Payment Income, MOU>
Where GovReceipt(t) = CBLoanInt(t) = |Project[<Interest earned, MOU>] (f[Central Bank Loan Interest](Central Bank)|

(6) Tax Transactions
(A) Tax for Company
Let ω ∈ Industry ∪ {Bank}.
f[Tax: ω](ω)
= CompTax[ω](t)<Taxes expense, MOU> + CompTax[ω](t)^<Cash, MOU>
f[Tax: ω](Government)
= CompTax[ω](t)<Taxes revenue, MOU> + CompTax[ω](t)<Cash, MOU>
For ω ∈ Industry
CompTax[ω](t)
= CompTaxRate(t) × |Project [<Gross Profit, MOU>]{¯f[Gross Profit: ω](ω)}|)
CompTax[Bank](t) = Max(0, CompTaxRate(t) × IncomInterest[Bank](t))
IncomInterest[Bank](t)
= |Project[<Interest earned, MOU>]{f[Loan Interest: ω](Bank)}|
− |Project[<Interest expense, MOU>]{f[Deposit Interest: ω](Bank)
+ f[Central Bank Loan Interest](Bank)}|
Where CompTaxRate(t) denotes the corporation tax rate. f[Gross Profit: ω](ω) is defined later.

(B) Household Income Tax
f[Tax: Household](Household)
= IncomeTax(t)<Taxes expense, MOU> + IncomeTax(t)^<Cash, MOU>
f[Tax: Household](Government)
= IncomeTax(t)<Taxes revenue, MOU> + IncomeTax(t)<Cash, MOU>

IncomeTax(t) = IncomeTaxRate(t) × Wages[Household](t)
Where IncomeTaxRate(t) denotes the income tax rate.

(C) Amount of Government Tax Revenue
The amount of Government tax revenue is given as follows.
TaxRevenue(t) = |∑{Project[<Taxes revenue, MOU>](f[Tax: ω](Government))
|ω ∈ {Bank} ∪ Industry}|
+ |Project[<Taxes revenue, MOU>](f[Tax: Household](Government))|
= ∑{CompTax[ω](t) | ω ∈ {Bank} ∪ Industry} + IncomeTax(t)

(6.3.3.6) [The Issue and Redemption of Central Bank Notes]

The Central Bank issues cash as a Central Bank note. At the same time the Central Bank also issues Central Bank note payable. If the Central Bank redeems cash (Central Bank note) then the cash is offset by a Central Bank note payable.

f[Central Bank Note Issue](Central Bank)
= IssueCash[Central Bank](t)<Central Bank note payable, MOU>
+ IssueCash[Central Bank](t)<Cash, MOU>
f[Central Bank Note Redeem](Central Bank)
= RedeemCash[Central Bank](t)^<Central Bank note payable, MOU>
+ RedeemCash[Central Bank](t)^<Cash, MOU>
IssueCash[Central Bank](t)
= |Project[^<Cash, MOU>](f[Reserve Fund Withdraw](Central Bank)
+ f[Central Bank Loan](Central Bank)
+ f[National Bond Central Bank Underwriting](Central Bank))|
RedeemCash[Central Bank](t)
= |Project[<Cash, MOU>](f[Reserve Fund Deposit](Central Bank)
+ f[Central Bank Repay](Central Bank)
+ f[National Bond Central Bank Redeem](Central Bank))|

(6.3.3.7) Transfer Transactions

(1) Gross Profit Transfer
Let ω ∈ Industry.
f[Transfer of Gross Profit: ω](ω)
= Wages[ω](t)^<Wages expenditure, MOU>
+ Wages[ω](t)^<Gross Profit, MOU>
+ Dep[ω](t)^<Depreciation, MOU>
+ Dep[ω](t)^<Gross Profit, MOU>
+ ValueAdd[ω](t)^<Value added, MOU>
+ ValueAdd[ω]<Gross Profit, MOU>
+ ValAdjustUp[ω](t)^<Value added, MOU>
+ ValAdjustUp[ω](t)<Gross Profit, MOU>
+ ValAdjustDown[ω](t)<Value added, MOU>
+ ValAdjustDown[ω](t)^<Gross Profit, MOU>

The transfer means that value added is transferred to the increase of gross profit and wages expenditure and depreciations are transferred as a decrease of gross profits. We also have to consider the value added adjustment factor. Then the following state of this economic field is defined.

f[Gross Profit: ω](ω)
= Project[^<Gross Profit, MOU>](f[Transfer of Gross Profit: ω](ω))
+ Project[<Gross Profit, MOU>](f[Transfer of Gross Profit: ω](ω))
f[Gross Profit Balance: ω](ω) = ¯{f[Gross Profit: ω](ω)}

(2) Ordinary Profit Transfers
Ordinary Profit Transfer is used for calculating ordinary profit from non-operating income and expenses such as interest earned, interest expenses, tax expenses and subsidies.

Let ω ∈ Industry. Let f[Gross Profit Balance: ω](ω)
= Sales Rev1[ω](t)<Gross Profit, MOU> + Sales Rev2[ω](t)^<Gross Profit, MOU>, where Sales Rev1[ω](t) = 0 or Sales Rev2[ω](t) = 0 from the definition of f[Gross Profit Balance: ω](ω).
f[Ordinary Profit Transfer: ω](ω)
= Sales Rev1[ω](t)^<Gross Profit, MOU>
+ Sales Rev1[ω](t)<Ordinary Profit, MOU>
+ Sales Rev2[ω](t)<Gross Profit, MOU>
+ Sales Rev2[ω](t)^<Ordinary Profit, MOU>
+ SaveInt[ω](t)^<Interest earned, MOU>
+ SaveInt[ω](t)<Ordinary Profit, MOU>
+ NBondInt[ω](t)^<Interest earned, MOU>
+ NBondInt[ω](t)<Ordinary Profit, MOU>
+ LoanInt[ω](t)^<Interest expense, MOU>
+ LoanInt[ω](t)^<Ordinary Profit, MOU>
+ Subsidy[ω](t)^<Subsidy Income, MOU>
+ Subsidy[ω](t)<Ordinary Profit, MOU>
+ CompTax[ω](t)^<Taxes expense, MOU>
+ CompTax[ω](t)^<Ordinary Profit, MOU>

Then ordinary profit is shown as follows.

f[Ordinary Profit: ω](ω)
= Project[^<Ordinary Profit, MOU>](f[Ordinary Profit Transfer: ω](ω))
+ Project[<Ordinary Profit, MOU>](f[Ordinary Profit Transfer: ω](ω))
f[Ordinary Profit Balance: ω](ω) = ¯{f[Ordinary Profit: ω](ω)}

(3) Retained Earnings Transfers
Ordinary Profit is transferred to retained earnings as follows.

Let f[Ordinary Profit Balance: ω](ω) = ¯{f[Ordinary Profit: ω](ω)}
= Revenue1[ω](t)<Ordinary Profit, MOU>

+ Revenue2[ω](t)^<Ordinary Profit, MOU>, where Revenue1[ω](t) = 0 or Revenue2[ω](t) = 0 from the definition of f[Ordinary Profit Balance: ω](ω).

Then retained earnings transfer is defined as follows.

f[Retained Earnings Transfer: ω](ω)
= Revenue1[ω](t)^<Ordinary Profit, MOU>
+ Revenue1[ω](t)<Retained Earnings, MOU>
+ Revenue2[ω](t)<Ordinary Profit, MOU>
+ Revenue2[ω](t)^<Retained Earnings, MOU>

f[Retained Earnings: ω](ω)
= Project[^<Retained Earnings, MOU>](f[Retained Earnings Transfer: ω](ω))
+ Project[<Retained Earnings, MOU>](f[Retained Earnings Transfer: ω](ω))
= Revenue1[ω](t)<Retained Earnings, MOU>
+ Revenue2[ω](t)^<Retained Earnings, MOU>

(4) Household: Ordinary Profit Transfers

In the case of households there is no transfer of gross profit. We only need ordinary profit transfer to calculate ordinary profit as follows.

f[Ordinary Profit Transfer: Household](Household)
= Wages[Household](t)^<Wages earned, MOU>
+ Wages[Household](t)<Ordinary Profit, MOU>
+ SaveInt[Household](t)^<Interest earned, MOU>
+ SaveInt[Household](t)<Ordinary Profit, MOU>
+ LoanInt[Household](t)^<Interest expense, MOU>
+ LoanInt[Household](t)^<Ordinary Profit, MOU>
+ Subsidy[Household](t)^<Subsidy Income, MOU>
+ Subsidy[Household](t)<Ordinary Profit, MOU>
+ NBondInt[Household](t)^<Interest earned, MOU>
+ NBondInt[Household](t)<Ordinary Profit, MOU>
+ IncomeTax(t)^<Taxes expense, MOU>
+ IncomeTax(t)^<Ordinary Profit, MOU>
+ HouseDomCons(t)^<Consumption expenditure, MOU>
+ HouseDomCons(t)^<Ordinary Profit, MOU>
+ Import[Bread](Household, t)^<Consumption expenditure, MOU>
+ Import[Bread](Household, t)^<Ordinary Profit, MOU>
+ Dep[Household](t)^<Depreciation, MOU>
+ Dep[Household](t)^<Ordinary Profit, MOU>

f[Ordinary Profit: Household](Household)
= Project[^<Ordinary Profit, MOU>](f[Ordinary Profit Transfer: Household](Household))
+ Project[<Ordinary Profit, MOU>](f[Ordinary Profit Transfer: Household](Household))
f[Ordinary Profit Balance: Household](Household)
= —{f[Ordinary Profit: Household](Household)}

(5) Household: Retained Earnings Transfers

Let f[Ordinary Profit Balance: Household](Household)
= Revenue1[Household](t)<Ordinary Profit, MOU>
+ Revenue2[Household](t)^<Ordinary Profit, MOU>.

f[Retained Earnings Transfer: Household](Household)
= Revenue1[Household](t)^<Ordinary Profit, MOU>
+ Revenue1[Household](t)<Retained Earnings, MOU>
+ Revenue2[Household](t)<Ordinary Profit, MOU>
+ Revenue2[Household](t)^<Retained Earnings, MOU>
f[Retained Earnings: Household](Household)
= Revenue1[Household](t)<Retained Earnings, MOU>
+ Revenue2[Household](t)^<Retained Earnings, MOU>

(6) Government: Ordinary Profit Transfers

f[Ordinary Profit Balance: Government](Government)
= TaxRevenue(t)^<Taxes revenue, MOU>
+ TaxRevenue(t)<Ordinary Profit, MOU>
+ GovReceipt(t)^<Central Bank Payment Income, MOU>
+ GovReceipt(t)<Ordinary Profit, MOU>
+ Dep[Government](t)^<Depreciation, MOU>
+ Dep[Government](t)^<Ordinary Profit, MOU>
+ Wages[Government](t)^<Wages expenditure, MOU>
+ Wages[Government](t)^<Ordinary Profit, MOU>
+ Subsidy[Government](t)^<Subsidy Expense, MOU>
+ Subsidy[Government](t)^<Ordinary Profit, MOU>
+ NBondInt[Government](t)^<Interest expense, MOU>
+ NBondInt[Government](t)^<Ordinary Profit, MOU>

Where,

TaxRevenue(t)
= ∑{CompTax[ω](t) | ω ∈ {Bank} ∪ Industry} + IncomeTax(t)
GovReceipt(t)
= |Project[<Central Bank Payment Income, MOU>](f[Central Bank Payment](Government))|
Dep[Government](t)
= |Project[<Depreciation, MOU>](f[Depreciation: Government](Government))|
Wages[Government](t)
= |Project[<Wages expenditure, MOU>](f[Employment: Government](Government))|
Subsidy[Government](t)
= |∑{Project[<Subsidy Expense, MOU>](f[Government Subsidy: ω](Government)) | ω ∈ Industry ∪ {Household, Bank}}|
NBondInt[Government](t)
= |∑{Project[<Interest expense, MOU>](f[National Bond Interest: ω](Government)) | ω ∈ {Household} ∪ Industry}|

f[Ordinary Profit: Government](Government)
= Project[^<Ordinary Profit, MOU>](f[Ordinary Profit Transfer: Government](Government))
+ Project[<Ordinary Profit, MOU>](f[Ordinary Profit Transfer: Government](Government))
f[Ordinary Profit Balance: Government](Government)
= —{f[Ordinary Profit: Government](Government)}

(7) Government: Retained Earnings Transfers
Let f[Ordinary Profit Balance: Government](Government)
= Revenue1[Government](t)<Ordinary Profit, MOU>
+ Revenue2[Government](t)^<Ordinary Profit, MOU>.

f[Retained Earnings Transfer: Government](Government)
= Revenue1[Government](t)^<Ordinary Profit, MOU>
+ Revenue1[Government](t)<Retained Earnings, MOU>
+ Revenue2[Government](t)<Ordinary Profit, MOU>
+ Revenue2[Government](t)^<Retained Earnings, MOU>
f[Retained Earnings: Government](Government)
= Revenue1[Government](t)<Retained Earnings, MOU>
+ Revenue2[Government](t)^<Retained Earnings, MOU>

(8) Bank: Ordinary Profit Transfers
f[Ordinary Profit Transfer: Bank](Bank)
= LoanIntIn[Bank](t)^<Interest earned, MOU>
+ LoanIntIn[Bank](t)<Ordinary Profit, MOU>
+ Subsidy[Bank](t)^<Subsidy Income, MOU>
+ Subsidy[Bank](t)<Ordinary Profit, MOU>
+ SaveIntOut[Bank](t)^<Interest expense, MOU>
+ SaveIntOut[Bank](t)^<Ordinary Profit, MOU>
+ LoanIntOut[Bank](t)^<Interest expense, MOU>
+ LoanIntOut[Bank](t)^<Ordinary Profit, MOU>
+ CompTax[Bank](t)^<Taxes expense, MOU>
+ CompTax[Bank](t)^<Ordinary Profit, MOU>

Where,

LoanIntIn[Bank]
= |Project[<Interest earned, MOU>](∑{f[Loan Interest Expense: ω](Bank) | ω ∈ Industry ∪ {Household}})|
Subsidy[Bank](t)
= |Project[<Subsidy Income, MOU>](f[Government Subsidy: Bank](Bank))|
SaveIntOut[Bank](t)
= |Project[<Interest expense, MOU>](∑{f[Deposit Interest: ω](Bank) | ω ∈ Industry ∪ {Household}})|
LoanIntOut[Bank](t)
= |Project[<Interest expense, MOU>](f[Central Bank Loan Interest Expense](Bank))| = CBLoanInt(t)

Then the ordinary profit of the bank is shown as follows.

f[Ordinary Profit: Bank](Bank)
= Project[^<Ordinary Profit, MOU>](f[Ordinary Profit Transfer: Bank]
(Bank))
+ Project[<Ordinary Profit, MOU>](f[Ordinary Profit Transfer: Bank](Bank))
f[Ordinary Profit Balance: Bank](Bank) = ¬{f[Ordinary Profit: Bank](Bank)}

(9) Bank: Retained Earnings Transfers
Let f[Ordinary Profit Balance: Bank](Bank)
= Revenue1[Bank](t)<Ordinary Profit, MOU>
+ Revenue2[Bank](t)^<Ordinary Profit, MOU>.

f[Retained Earnings Transfer: Bank](Bank)
= Revenue1[Bank](t)^<Ordinary Profit, MOU>
+ Revenue1[Bank](t)<Retained Earnings, MOU>
+ Revenue2[Bank](t)<Ordinary Profit, MOU>
+ Revenue2[Bank](t)^<Retained Earnings, MOU>
f[Retained Earnings: Bank](Bank)
= Revenue1[Bank](t)<Retained Earnings, MOU>
+ Revenue2[Bank](t)^<Retained Earnings, MOU>

(10) Central Bank: Ordinary Profit Transfers
There are no retained earnings in the Central Bank. Ordinary profit is always zero from the following definition.

f[Ordinary Profit Transfer: Central Bank](Central Bank)
= CBLoanInt(t)^<Interest earned, MOU>
+ CBLoanInt(t)<Ordinary Profit, MOU>
+ GovReceipt(t)^<Central Bank Payment Expense, MOU>
+ GovReceipt(t)^<Ordinary Profit, MOU>

Where,

CBLoanInt(t)
= |Project[<Interest earned, MOU>](f[Central Bank Loan Interest Expense]
(Central Bank))|
GovReceipt(t) = CBLoanInt(t)

(11) Interim Flow
(11-1) Industry: Interim Flow
Let ω ∈ Industry.
f[Interim flow](ω, t)
= Flow[Cash](ω, t)<Cash, MOU>
+ Flow[^Cash](ω, t)^<Cash, MOU>
+ Flow[Retained Earnings](ω, t)<Retained Earnings, MOU>
+ Flow[^Retained Earnings](ω, t)^<Retained Earnings, MOU>
+ Flow[Deposits](ω, t)<Deposits, MOU>
+ Flow[^Deposits](ω, t)^<Deposits, MOU>

+ Flow[National Bonds](ω, t)<National Bonds, MOU>
+ Flow[^National Bonds](ω, t)^<National Bonds, MOU>
+ Flow[Loans payable](ω, t)<Loans payable, MOU>
+ Flow[^Loans payable](ω, t)^<Loans payable, MOU>
+ Flow[em(ω)](ω, t)<em(ω), MOU>
+ Flow[^em(ω)](ω, t)^<em(ω), MOU>
+ Flow[ep(ω)](ω, t)<ep(ω), MOU>
+ Flow[^ep(ω)](ω, t)^<ep(ω), MOU>
+ Flow[Machinery]((ω, t)<Machinery, MOU>
+ Flow[^Machinery]((ω, t)<Machinery, MOU>
+ Flow[Reserve for depreciation](ω, t)<Reserve for depreciation, MOU>
+ Flow[^Reserve for depreciation](ω, t)^<Reserve for depreciation, MOU>

Where ep(ω) = WhatProduct(ω), em(ω) = WhatMaterial(ep(ω))

Agriculture and the Steel Industry do not have Flow[em(ω)](ω, t)<em(ω), MOU> and Flow[^em(ω)](ω, t)^<em(ω), MOU> because there are no materials used in these industries.

Where,

Flow[Cash](ω, t)
= |Project[<Cash, MOU>]{f[Sell: ep(ω): α](ω)
+ f[Export: e](ω)
+ f[Withdraw Deposit: ω](ω)
+ f[National Bond Redeem: ω](ω)
+ f[Loan: ω](ω)
+ f[Government Subsidy: ω](ω)
+ f[Deposit Interest Transaction: ω](ω)
+ f[National Bond Interest: ω](ω)}|

Flow[^Cash](ω, t)
= |Project[^<Cash, MOU>]{f[Buy: em(ω): α](ω)
+ f[Import: e](ω)
+ f[Employment: ω](ω)
+ f[Repay: ω](ω)
+ f[Deposit: ω](ω)
+ f[Loan Interest: ω](ω)
+ f[Tax: ω](ω)
+ f[National Bond Purchase: ω](ω)
+ f[Domestic Equipment investments: Machinery: ω](ω)}|

Flow[Retained Earnings](ω, t)
= |Project[<Retained Earnings, MOU>](f[Retained Earnings: ω](ω))|

Flow[^Retained Earnings](ω, t)
= |Project[^<Retained Earnings, MOU>](f[Retained Earnings: ω](ω))|

Flow[Deposits](ω, t)
= |Project[<Deposits, MOU>](f[Deposit: ω](ω))|

Flow[^Deposits](ω, t)
= |Project[^<Deposits, MOU>](f[Withdraw Deposit: ω](ω))|

Flow[National Bonds](ω, t)
= |Project[<National Bonds, MOU>](f[National Bond Purchase: ω](ω))|

Flow[^National Bonds](ω, t)
= |Project[^<National Bonds, MOU>](f[National Bond Redeem: ω](ω))|

Flow[Loans payable](ω, t)
= |Project[<Loans payable, MOU>](f[Loan: ω](ω))|

Flow[^Loans payable](ω, t)
= |Project[^<Loans payable, MOU>](f[Repay: ω](ω))|

Flow[em(ω)](ω, t)
= |Project[<em(ω), MOU>](f[Buy: em(ω): α](ω)
+ f[Import: em(ω)](ω))|

Flow[^em(ω)](ω, t)
= |Project[^<em(ω), MOU>](f[Production: ep(ω)](ω))|

Flow[ep(ω)](ω, t)
= |Project[<ep(ω), MOU>](f[Production: ep(ω)](ω))|

Flow[^ep(ω)](ω, t)
= |Project[^<ep(ω), MOU>](f[Sell: ep(ω): α](ω)
+ f[Export: ep(ω)](ω))|

Flow[Reserve for depreciation](ω, t)
= |Project[<Reserve for depreciation, MOU>](f[Depreciation: ω](ω))|

Flow[^Reserve for depreciation](ω, t)
= |Project[^<Reserve for depreciation, MOU>](f[Scrap of Equipment: ω](ω))|

Flow[Machinery]((ω, t)
= |Project[<Machinery, MOU>](f[Domestic Equipment investments: Machinery: ω](ω)
+ f[Import: Machinery](Household))|

Flow[^Machinery]((ω, t)
= |Project[^<Machinery, MOU>](f[Scrap of Equipment: ω](ω))|

(11-2) Households: Interim Flow
f[Interim flow](Household, t)
= Flow[Cash](Household, t)<Cash, MOU>
+ Flow[^Cash](Household, t)^<Cash, MOU>
+ Flow[Retained Earnings](Household, t)<Retained Earnings, MOU>
+ Flow[^Retained Earnings](Household, t)^<Retained Earnings, MOU>
+ Flow[Deposits](Household, t)<Deposits, MOU>
+ Flow[^Deposits](Household, t)^<Deposits, MOU>

+ Flow[National Bonds](Household, t)<National Bonds, MOU>
+ Flow[^National Bonds](Household, t)^<National Bonds, MOU>
+ Flow[Loans payable](Household, t)<Loans payable, MOU>
+ Flow[^Loans payable](Household, t)^<Loans payable, MOU>
+ Flow[Machinery]((Household, t)<Machinery, MOU>
+ Flow[^Machinery]((Household, t)<Machinery, MOU>
+ Flow[Reserve for depreciation](Household, t)
<Reserve for depreciation, MOU>
+ Flow[^Reserve for depreciation](Household, t)
^<Reserve for depreciation, MOU>

Where,

Flow[Cash](Household, t)
= |Project[<Cash, MOU>]{f[Employment: ω](Household)
+ f[Withdraw Deposit: Household](Household)
+ f[National Bond Redeem: Household](Household)
+ f[Loan: Household](Household)
+ f[Government Subsidy: Household](Household)
+ f[Deposit Interest: Household](Household)
+ f[National Bond Interest: Household](Household)}|

Flow[^Cash](Household, t)
= |Project[^<Cash, MOU>]{f[Buy: Bread: Bread Industry](Household)
+ f[Import: Bread](Household)
+ f[Import: Machinery](Household)
+ f[Repay: Household](Household)
+ f[Deposit: Household](Household)
+ f[Loan Interest Expense: Household](Household)
+ f[Tax: Household](Household)
+ f[National Bond Purchase: Household](Household)
+ f[Domestic Equipment investments: Machinery: Household](Household)}|

Flow[Retained Earnings](Household, t)
= |Project[<Retained Earnings, MOU>](f[Retained Earnings: Household](Household))|

Flow[^Retained Earnings](Household, t)
= |Project[^<Retained Earnings, MOU>](f[Retained Earnings: Household](Household))|

Flow[Deposits](Household, t)
= |Project[<Deposits, MOU>](f[Deposit: Household](Household))|

Flow[^Deposits](Household, t)
= |Project[^<Deposits, MOU>](f[Withdraw Deposit: Household](Household))|

Flow[National Bonds](Household, t)
= |Project[<National Bonds, MOU>](f[National Bond Purchase: Household](Household))|

Flow[^National Bonds](Household, t)
= |Project[^<National Bonds, MOU>](f[National Bond Redeem: Household](Household))|

Flow[Loans payable](Household, t)
= |Project[<Loans payable, MOU>](f[Loan: Household](Household))|

Flow[^Loans payable](Household, t)
= |Project[^<Loans payable, MOU>](f[Repay: Household](Household))|

Flow[Machinery]((Household, t)
= |Project[<Machinery, MOU>](f[Domestic Equipment investments: Machinery: Household](Household)
+ f[Import: Machinery](Household))|

Flow[^Machinery]((Household, t)
= |Project[^<Machinery, MOU>](f[Scrap of Equipment: Household](Household))|

Flow[Reserve for depreciation](Household, t)
= |Project[<Reserve for depreciation, MOU>](f[Depreciation: Household](Household))|

Flow[^Reserve for depreciation](Household, t)
= |Project[^<Reserve for depreciation, MOU>]
(f[Scrap of Equipment: Household](Household))|

(11-3) Government: Interim Flow
f[Interim flow](Government, t)
= Flow[Cash](Government, t)<Cash, MOU>
+ Flow[^Cash](Government, t)^<Cash, MOU>
+ Flow[Retained Earnings](Government, t)<Retained Earnings, MOU>
+ Flow[^Retained Earnings](Government, t)^<Retained Earnings, MOU>
+ ∑{Flow[National Bonds payable: ω](Government, t)
<National Bonds payable, MOU> | ω ∈ {Household} ∪ Industry}
+ ∑{Flow[^National Bonds payable: ω](Government, t)
^<National Bonds payable, MOU> | ω ∈ {Household} ∪ Industry}
+ Flow[National Bonds payable: Central Bank](Government, t)
<National Bonds payable, MOU>
+ Flow[^National Bonds payable: Central Bank](Government, t)
^<National Bonds payable, MOU>
+ Flow[Machinery]((Government, t)<Machinery, MOU>
+ Flow[^Machinery]((Government, t)<Machinery, MOU>
+ Flow[Reserve for depreciation](Government, t)
<Reserve for depreciation, MOU>
+ Flow[^Reserve for depreciation](Government, t)
^<Reserve for depreciation, MOU>

Where,

Flow[Cash](Government, t)
= |Project[<Cash, MOU>]{f[Central Bank Payment](Government)
+ ∑{f[Tax: ω](Government) | ω ∈ {Household, Bank} ∪ Industry}
+ f[National Bond Purchase: ω](Government)
+ f[National Bond Central Bank Underwriting](Government)}|

Flow[^Cash](Government, t)
= |Project[^<Cash, MOU>]{∑{f[National Bond Redeem: ω](Government) | ω ∈ Industry ∪ {Household}}
+ f[National Bond Central Bank Redeem](Government)
+ f[Employment: Government](Government)
+ ∑{f[Government Subsidy: ω](Government) | ω ∈ Industry ∪ {Household, Bank}}
+ ∑{f[National Bond Interest: ω](Government) | ω ∈ Industry ∪ {Household}}
+ f[Domestic Equipment investments: Machinery: Government](Government)}
+ f[Import: Machinery](Government)|

Flow[Retained Earnings](Government, t)
= |Project[<Retained Earnings, MOU>](f[Retained Earnings: Government](Government))|

Flow[^Retained Earnings](Government, t)
= |Project[^<Retained Earnings, MOU>](f[Retained Earnings: Government](Government))|

Let ω ∈ Industry ∪ {Household}.
Flow[National Bonds payable: ω](Government, t)
= |Project[<National Bonds payable, MOU>]
(f[National Bond Purchase: ω](Government))|

Flow[^National Bonds payable: ω](Government, t)
= |Project[^<National Bonds payable, MOU>]
(f[National Bond Redeem: ω](Government))|

Flow[National Bonds payable: Central Bank](Government, t)
= |Project[<National Bonds payable, MOU>]
(f[National Bond Central Bank Underwriting](Government)|

Flow[^National Bonds payable: Central Bank](Government, t)
= |Project[^<National Bonds payable, MOU>]
(f[National Bond Central Bank Redeem](Government))|

Flow[Machinery]((Government, t)
= |Project[<Machinery, MOU>]
(f[Domestic Equipment investments: Machinery: Government](Government)
+ f[Import: Machinery](Government))|

Flow[^Machinery]((Government, t)
= |Project[^<Machinery, MOU>](f[Scrap of Equipment: Government]
(Government))|

Flow[Reserve for depreciation](Government, t)
= |Project[<Reserve for depreciation, MOU>](f[Depreciation: Government]
(Government))|

Flow[^Reserve for depreciation](Government, t)
= |Project[^<Reserve for depreciation, MOU>]
(f[Scrap of Equipment: Government](Government))|

(11-4) Bank: Interim Flow
f[Interim flow](Bank, t)
= Flow[Cash](Bank, t)<Cash, MOU>
+ Flow[^Cash](Bank, t)^<Cash, MOU>
+ ∑{Flow[Deposits payable: ω](Bank, t)<Deposits payable, MOU> | ω ∈
{Household} ∪ Industry}
+ ∑{Flow[^Deposits payable: ω](Bank, t)^<Deposits payable, MOU> | ω ∈
{Household} ∪ Industry}
+ ∑{Flow[Loans receivable: ω](Bank, t)<Loans receivable, MOU> | ω ∈
{Household} ∪ Industry}
+ ∑{Flow[^Loans receivable: ω](Bank, t)^<Loans receivable, MOU> | ω ∈
{Household} ∪ Industry}
+ Flow[Central Bank Reserve Fund Receivable](Bank, t)
<Reserve deposit receivable, MOU>
+ Flow[^Central Bank Reserve Fund Receivable](Bank, t)
^<Reserve deposit receivable, MOU>
+ Flow[Central BankLoans payable](Bank, t)<Loans payable, MOU>
+ Flow[^Central BankLoans payable](Bank, t)^<Loans payable, MOU>
+ Flow[Retained Earnings](Bank, t)<Retained Earnings, MOU>
+ Flow[^Retained Earnings](Bank, t)^<Retained Earnings, MOU>

Where,

Flow[Cash](Bank, t)
= |Project[<Cash, MOU>]{∑{f[Repay: ω](Bank) | ω ∈ Industry ∪ {Household}}
+ ∑{f[Deposit: ω](Bank) | ω ∈ Industry ∪ {Household}}
+ f[Central Bank Loan](Bank)
+ f[Reserve Fund Withdraw](Bank)
+ f[Government Subsidy: Bank](Bank)
+ ∑{f[Loan Interest Expense: ω](Bank) | ω ∈ Industry ∪ {Household}}}|

Flow[^Cash](Bank, t)
= |Project[^<Cash, MOU>]{∑{f[Loan: ω](Bank) | ω ∈ Industry ∪ {Household}}
+ ∑{f[Withdraw Deposit: ω](Bank) | ω ∈ Industry ∪ {Household}}
+ f[Central Bank Repay](Bank)
+ f[Reserve Fund Deposit](Bank)

+ Σ{f[Deposit Interest: ω](Bank) | ω ∈ Industry ∪ {Household}}
+ f[Central Bank Loan Interest Expense](Bank)
+ f[Tax: Bank](Bank)}|

Flow[Deposits payable: ω](Bank, t)
= |Project[<Deposits payable, MOU>](Σ{f[Deposit: ω](Bank) | ω ∈ Industry ∪ {Household}})|

Flow[^Deposits payable: ω](Bank, t)
= |Project[^<Deposits payable, MOU>](Σ{f[Withdraw Deposit: ω](Bank) | ω ∈ Industry ∪ {Household}})|

Flow[Loans receivable: ω](Bank, t)
= |Project[<Loans receivable, MOU>](Σ{f[Loan: ω](Bank) | ω ∈ Industry ∪ {Household}})|

Flow[^Loans receivable: ω](Bank, t)
= |Project[^<Loans receivable, MOU>](Σ{f[Repay: ω](Bank) | ω ∈ Industry ∪ {Household}})|

Flow[Central Bank Reserve Fund Receivable](Bank, t)
= |Project[<Reserve deposit receivable, MOU>](f[Reserve Fund Deposit](Bank))|

Flow[^Central Bank Reserve Fund Receivable](Bank, t)
= |Project[^<Reserve deposit receivable, MOU>](f[Reserve Fund Withdraw](Bank))|

Flow[Central BankLoans payable](Bank, t)
= |Project[<Loans payable, MOU>](f[Central Bank Loan](Bank))|

Flow[^Central BankLoans payable](Bank, t)
= |Project[^<Loans payable, MOU>](f[Central Bank Repay](Bank))|

Flow[Retained Earnings](Bank, t)
= |Project[<Retained Earnings, MOU>](f[Retained Earnings: Bank](Bank))|

Flow[^Retained Earnings](Bank, t)
= |Project[^<Retained Earnings, MOU>](f[Retained Earnings: Bank](Bank))|

(11-5) Central Bank: Interim Flow
f[Interim flow](Central Bank, t) =
Flow[Central BankLoans receivable](Central Bank, t)
<Loans receivable, MOU>
+ Flow[^Central BankLoans receivable](Central Bank, t)
^<Loans receivable, MOU>
+ Flow[National Bonds](Central Bank, t)<National Bonds, MOU>
+ Flow[^National Bonds](Central Bank, t)^<National Bonds, MOU>
+ Flow[Central Bank Reserve Fund Payable](Central Bank, t)
<Reserve deposit payable, MOU>
+ Flow[^Central Bank Reserve Fund Payable](Central Bank, t)
^<Reserve deposit payable, MOU>

+ Flow[Central Bank note payable](Central Bank, t)
<Central Bank note payable, MOU>
+ Flow[^Central Bank note payable](Central Bank, t)
^<Central Bank note payable, MOU>

Where,

Flow[Central BankLoans receivable](Central Bank, t)
= |Project[<Loans receivable, MOU>](f[Central Bank Loan](Central Bank))|

Flow[^Central BankLoans receivable](Central Bank, t)
= |Project[^<Loans receivable, MOU>](f[Central Bank Repay]
(Central Bank))|

Flow[National Bonds](Central Bank, t)
= |Project[<National Bonds, MOU>]
(f[National Bond Central Bank Underwriting](Central Bank))|

Flow[^National Bonds](Central Bank, t)
= |Project[^<National Bonds, MOU>]
(f[National Bond Central Bank Redeem](Central Bank))|

Flow[Central Bank Reserve Fund Payable](Central Bank, t)
= |Project[<Reserve deposit payable, MOU>](f[Reserve Fund Deposit]
(Central Bank))|

Flow[^Central Bank Reserve Fund Payable](Central Bank, t)
= |Project[^<Reserve deposit payable, MOU>](f[Reserve Fund Withdraw]
(Central Bank))|

Flow[Central Bank note payable](Central Bank, t)
= |Project[<Central Bank note payable, MOU>](f[Central Bank Note Issue]
(Central Bank))|

Flow[^Central Bank note payable](Central Bank, t)
= |Project[^<Central Bank note payable, MOU>](f[Central Bank Note Redeem]
(Central Bank))|

(12) Final Stock
f[Final Stock](ω, t) = ¯{f[Initial Stock](ω, t) + f[Interim flow](ω, t)}
Where, ω ∈ Industry ∪ {Bank, Central Bank, Household, Government}

6.3.4 Bottom Up Construction of Macro Economic Variables

(6.3.4.1) [Sector-Deposit: S(ω)]

Let ω ∈ Industry ∪ {Household, Government}.
S(ω) = S[Retained Earnings](ω) + S[Depreciation](ω)
S(Bank) = S[Retained Earnings](Bank)

(1)
Let ω ∈ Industry.
S[Retained Earnings](ω)
= |Project[<Retained Earnings, MOU>](f[Retained Earnings: ω](ω))|
– |Project[^<Retained Earnings, MOU>](f[Retained Earnings: ω](ω))|
= Revenue1[ω](t) – Revenue2[ω](t)

Where,

Revenue1[ω](t)
= Sales Rev1[ω](t) + SaveInt[ω](t) + NBondInt[ω](t) + Subsidy[ω](t)
Revenue2[ω](t) = Sales Rev2[ω](t) + LoanInt[ω](t) + CompTax[ω](t)
Sales Rev1[ω](t) = ValueAdd[ω] + ValAdjustUp[ω](t)
Sales Rev2[ω](t) = Wages[ω](t) + Dep[ω](t) + ValAdjustDown[ω](t)

(2)
Let ω = Household.
S[Retained Earnings](Household)
= Revenue1[Household](t) – Revenue2[Household](t)

Where,

Revenue1[Household](t)
= Wages[Household](t) + SaveInt[Household](t) + NBondInt[Household](t)
+ Subsidy[Household](t)
Revenue2[Household](t)
= LoanInt[Household](t) + IncomeTax(t) + HouseDomCons(t)
+ Import[Bread](Household, t) + Dep[Household](t)

(3)
Let ω = Government.
S[Retained Earnings](Government)
= Revenue1[Government](t) – Revenue2[Government](t)

Where,

Revenue1[Government](t) = TaxRevenue(t) + GovReceipt(t)
Revenue2[Government](t)
= Dep[Government](t) + Wages[Government](t) + Subsidy[Government](t)
+ NBondInt[Government](t)

(4)
Let ω = Bank.
S[Retained Earnings](Bank) = Revenue1[Bank](t) – Revenue2[Bank](t)

Where,

Revenue1[Bank](t) = LoanIntIn[Bank](t) + Subsidy[Bank](t)
Revenue2[Bank](t)
= SaveIntOut[Bank](t) + LoanIntOut[Bank](t) + CompTax[Bank](t)

(5)
Let ω ≠ Bank.
S[Depreciation](ω)
= |Project[<Reserve for depreciation, MOU>](f[Depreciation: ω](ω))|

(6.3.4.2) [Sector-Investment: I(ω)]

(1)
Let ω ∈ {Flour-milling Industry, Bread Industry, Machinery Industry}.
I(ω) = I[Equipment investments](ω)
+ I[Product Stock Investment](ω)
+ I[Material Stock Investment](ω)
+ I[Value added_Product Stock Adjustment](ω)

(2)
Let ω ∈ {Steel Industry, Agriculture}.
I(ω) = I[Equipment investments](ω)
+ I[Product Stock Investment](ω)
+ I[Value added_Product Stock Adjustment](ω)

(3)
Let ω ∈ {Government, Household}.
I(ω) = I[Equipment investments](ω)

(A) Equipment investments: I[Equipment investments](ω)
Let ep(ω) = WhatProduct(ω) and ω ∈ Industry ∪ {Government, Household}.
I[Equipment investments](ω)
= |Project[<Machinery, MOU>](f[Domestic Equipment investments: Machinery: ω](ω) + f[Import: Machinery](ω)) |
= InvPrice[ω](t) + Import[Machinery](ω, t)

(B) Product Stock Investments: I[Product Stock Investment: ep(ω)](ω)
Let ω ∈ {Agriculture, Steel Industry} and α buys the product of ω.
I[Product Stock Investment: ep(ω)](ω)
= |Project[<ep(ω), MOU>](f[Production: ep(ω)](ω))|
– |Project[<^ep(ω), MOU>](f[Sell: ep(ω): α](ω) + f[Export: ep(ω)](ω))|
= ProPrice[ep(ω)](t) – SellPrice[ep(ω)](t) – Export[ep(ω)](ω, t)
= ValueAdd[ω](t) – SellPrice[ep(ω)](t) – Export[ep(ω)](ω, t)

Let ω ∈ {Flour-milling Industry, Bread Industry} and α buys the product of ω.
I[Product Stock Investment: ep(ω)](ω)
= |Project[<ep(ω), MOU>](f[Production: ep(ω)](ω))|
– |Project[<^ep(ω), MOU>](f[Sell: ep(ω): α](ω) + f[Export: ep(ω)](ω))|
= ProPrice[ep(ω)](t) – SellPrice[ep(ω)](t) – Export[ep(ω)](ω, t)
= Use[em(ω)](t) + ValueAdd[ω](t) – SellPrice[ep(ω)](t) – Export[ep(ω)](ω, t)

Let ω ∈ {Machinery Industry}.
I[Product Stock Investment: Machinery](Machinery Industry)
= |Project[<Machinery, MOU>](f[Production: Machinery](Machinery Industry))|

– |Project[<^Machinery, MOU>] (∑{f[Sell: Machinery: ω]
(Machinery Industry) | ω ∈ Industry ∪ {Household, Government}}
+ f[Export: Machinery](Machinery Industry))|
= ProPrice[Machinery](t)
– ∑{InvPrice[ω](t) | ω ∈ Industry ∪ {Household, Government}}
– Export[Machinery](Machinery Industry, t)
= Use[Steel](t) + ValueAdd[Machinery Industry](t)
– ∑{InvPrice[ω](t) | ω ∈ Industry ∪ {Household, Government}}
– Export[Machinery](Machinery Industry, t)

(C) Material Stock Investments: I[Material Stock Investment: em(ω)](ω)
Let ω ∈ {Flour-milling Industry, Bread Industry, Machinery Industry} and ω purchases the material from β.
I[Material Stock Investment: em(ω)](ω)
= |Project[<em(ω), MOU>](f[Buy: em(ω): β](ω)
+ f[Import: em(ω)](ω))|
– |Project[^<em(ω), MOU>](f[Production: ep(ω)](ω))|
= SellPrice[em(ω)](t) + Import[em(ω)](ω, t) – Use[em(ω)](t)

(D) Value added_Product Stock Adjustment:
I[Value added_Product Stock Adjustment](ω)
Let ω ∈ Industry.
I[Value added_Product Stock Adjustment](ω)
= |Project[<ep(ω), MOU>](f[Value added_Product Stock Adjustment](ω))|
– |Project[^<ep(ω), MOU>](f[Value added_Product Stock Adjustment](ω))|
= ValadjustUp[ω](t) – ValAdjustDown[ω](t)

(E) Stock Investments: Transfer Transactions
Transfer transactions are defined as follows, that are convention for calculation.

(1) Product Stock Investment Transfers:
f[Product Stock Investment Transfer: ep(ω)](ω)
Let ω ∈ Industry.
f[Product Stock Investment Transfer: ep(ω)](ω)
= Max(0, I[Product Stock Investment: ep(ω)](ω))^<ep(ω), MOU>
+ Max(0, I[Product Stock Investment: ep(ω)](ω))<Stock Investment, MOU>
+ Max(0, –I[Product Stock Investment: ep(ω)](ω))<ep(ω), MOU>
+ Max(0, –I[Product Stock Investment: ep(ω)](ω))^<Stock Investment, MOU>

(2) Material Stock Investment Transfers:
f[Material Stock Investment Transfer: em(ω)](ω)
Let ω ∈ {Flour-milling Industry, Bread Industry, Machinery Industry} and ω purchases material from β.
f[Material Stock Investment Transfer: em(ω)](ω)
= Max(0, I[Material Stock Investment: em(ω)](ω))^<em(ω), MOU>
+ Max(0, I[Material Stock Investment: em(ω)](ω))<Stock Investment, MOU>

+ Max(0, –I[Material Stock Investment: em(ω)](ω))<em(ω), MOU>
+ Max(0, –I[Material Stock Investment: em(ω)](ω))^<Stock Investment, MOU>

(F) Equipment investments: Transfer Transaction:
f[Equipment investments Transfer: Machinery](ω)
Let ω ∈ Industry ∪ {Government, Household}.

Transfer transaction of equipment investments is defined as follows. This transfer is a convention for calculation.

f[Equipment investments Transfer: Machinery](ω)
= I[Equipment investments](ω)^<Machinery, MOU>
+ I[Equipment investments](ω)<Equipment investments, MOU>

(6.3.4.3) [Sector-Value added: Y(ω)]

(1) Let ω ∈ Industry
Y(ω) = |Project[<Value added, MOU>](f[Production: ep(ω)](ω)
+ f[Value added_Product Stock Adjustment](ω))|
– |Project[^<Value added, MOU>](f[Value added_Product Stock Adjustment](ω))|
= ValueAdd[ω](t) + ValAdjustUp[ω](t) – ValAdjustDown[ω](t)

(2) Y(Government)
Y(Government)
= |Project[<Value added, MOU>](f[Government Service Production](Government))|
= Wages[Government](t)

(6.3.4.4) [Sector-Consumption: C(ω)]

C(Household)
= |Project[<Consumption expenditure, MOU>](f[Household_Consumption expenditure](Household))
= HouseDomCons(t) + Import[Bread](Household, t)
C(Government)
= |Project[<Consumption expenditure, MOU>](f[Government Service Expense](Government))|
= Wages[Government](t)

(6.3.4.5) [Sector-Export & Import: EX[e](ω), IM[e](ω)]

Let ep(ω) = WhatProduct(ω), em(ω) = WhatMaterial(ep(ω)).

(1) EX[ep(ω)](ω), ω ∈ Industry
EX[ep(ω)](ω)
= |Project[^<ep(ω), MOU>](f[Export: ep(ω)](ω))|
= Export[ep(ω)](ω, t)

(2a) IM[em(ω)](ω), ω ∈ {Flour-milling Industry, Bread Industry, Machinery Industry}
IM[em(ω)](ω)
= |Project[<em(ω), MOU>](f[Import: em(ω)](ω))|
= Import[em(ω)](ω, t)

(2b) IM[Machinery](ω), ω ∈ Industry ∪ {Government, Household}
IM[Machinery](ω)
= |Project[<Machinery, MOU>](f[Import: Machinery](ω))|
= Import[Machinery](ω, t)

(3) IM[Bread](Household), ω = Household
IM[Bread](Household)
= |Project[<Bread, MOU>](f[Import: Bread](Household))|
= Import[Bread](Household, t)

(6.3.4.6) [Macro National Economic Variables]

S[National Economy]
= Σ{S(ω) | ω ∈ Industry ∪ {Household, Government, Bank}}
I[National Economy] = Σ{I(ω) | ω ∈ Industry ∪ {Household, Government}}
Y[National Economy] = Σ{Y(ω) | ω ∈ Industry ∪ {Government}}
C[National Economy] = C(Household) + C(Government)
EX[National Economy] = Σ{EX[ep(ω)](ω) | ω ∈ Industry}
IM[National Economy]
= Σ{IM[em(ω)](ω) | ω ∈ {Flour-milling Industry, Bread Industry, Machinery Industry}}
+ Σ{IM[Machinery](ω) | ω ∈ Industry ∪ {Government, Household}
+ IM[Bread](Household)}

Notice that Y(ω) = S(ω) + C(ω) and Y(ω) = I(ω) + C(ω) + EX[ep(ω)](ω) − IM[em(ω)](ω) − IM[Machinery](ω) do not hold in each sector level. On the other hand (1) Y[National Economy] = S[National Economy] + C[National Economy] and (2) Y[National Economy] = I[National Economy] + C[National Economy] + EX[National Economy] − IM[National Economy] hold in macro economy level as well known identity equations. But the equations require the proof. We give the proof in the following section 6.6.3.

6.4 National Economic Field: Aggregation

[Definition 6.4.1] Basic Set of Economic Units Ω_0, Basic Account Titles Λ_0
Ω_0 = {Agriculture, Flour-milling industry, Bread Industry, Steel Industry, Machinery Industry, Household, Government, Bank, Central Bank, Overseas}
Where {Agriculture, Flour-milling Industry, Bread Industry, Steel Industry, Machinery Industry} = Industry, {Bank, Central Bank} = Banking Enterprises

[Definition 6.4.2] Sector-Set Ω_1, Integrated-Set Ω_2, Map g_1, g_2
Let Ω_1 = {Industry, Household, Government, Bank, Central Bank, Oversea} and Ω_2 = {Nation, Oversea}.
Aggregation Map $g_1: \Omega_0 \to \Omega_1$, $g_2: \Omega_1 \to \Omega_2$ are defined as follows.

(1) $g_1: \Omega_0 \to \Omega_1$
(a) If $\omega \in$ Industry then $g_1(\omega)$ = Industry
(b) If $\omega \in$ {Household, Bank, Central Bank, Government, Oversea} then $g_1(\omega) = \omega$

(2) $g_2: \Omega_1 \to \Omega_2$
(a) If $\omega \in$ {Industry, Household, Bank, Central Bank, Government} then $g_2(\omega)$ = Nation
(b) If $\omega \in$ {Oversea} then $g_2(\omega) = \omega$

[Definition 6.4.3] Basic National Economic Field, Sector-National Economic Field, Integrated National Economic Field
Basic National Economic Field is denoted by $F[\Omega_0, \Gamma_0]$, Sector-National Economic Field is denoted by $F(\Omega_1, \Gamma_0)$ and Integrated National Economic Field is denoted by $F[\Omega_2, \Gamma_0]$.
Where Γ_0 denotes Γ[Vintual Economy] or Γ[Virtual Economy: Multi Dimension].

[Example 6.4.1] National Economic Field: Initial Stock
Let $F[\Omega_0, \Gamma_0] = \{f \mid f: \Omega_0 \to \Gamma_0\}$.
f[Initial Stock]$(\omega, t) \in F[\Omega_0, \Gamma_0]$ is given in this example as follows.

[Initial Stock]
f[Initial Stock](Agriculture, t) = 150<Cash, MOU>
+ 50<Machinery, MOU> + 100<Capital stock, MOU>
+ 100<Retained Earnings, MOU>
f[Initial Stock](Flour-milling Industry, t) = 150<Cash, MOU>
+ 250<Machinery, MOU> + 300<Capital stock, MOU>
+ 100<Retained Earnings, MOU>
f[Initial Stock](Bread Industry, t) = 350<Cash, MOU>
+ 150<Machinery, MOU> + 200<Capital stock, MOU>
+ 300<Retained Earnings, MOU>
f[Initial Stock](Steel Industry, t) = 100<Cash, MOU>
+ 100<Machinery, MOU> + 100<Capital stock, MOU>
+ 100<Retained Earnings, MOU>
f[Initial Stock](Machinery Industry, t) = 150<Cash, MOU>
+ 250<Machinery, MOU> + 250<Capital stock, MOU>
+ 150<Retained Earnings, MOU>
f[Initial Stock](Household, t) = 500<Cash, MOU> + 500<Machinery, MOU>
+ 400<Capital stock, MOU> + 600<Retained Earnings, MOU>
f[Initial Stock](Government, t) = 500<Cash, MOU> + 150<Machinery, MOU>
+ 200<Capital stock, MOU> + 300<Retained Earnings, MOU>

+ 150<National Bonds payable, MOU>
f[Initial Stock](Bank, t) = 500<Cash, MOU> + 50<Machinery, MOU>
+ 200<Capital stock, MOU> + 300<Retained Earnings, MOU>
+ 300<Reserve deposit receivable, MOU> + 350<Loans payable, MOU>
f[Initial Stock](Central Bank, t) = 0<Cash, MOU>
+ 1000<Capital stock, MOU> + 300<Reserve deposit payable, MOU>
+ 2400<Central Bank note payable, MOU> + 350<Loans receivable, MOU>
+ 3200<Gold, MOU> + 150<National Bonds, MOU>

f[Initial Stock](Oversea, t) = 0

Note: Cash in the Central Bank is setoff by a Central Bank note payable.

[Definition 6.4.4] Simple-Sector-Aggregation Map
Simple-Sector-Aggregation map: $G_0: F(\Omega_0, \Gamma_0) \rightarrow F(\Omega_1, \Gamma_0)$ is defined for $f \in F(\Omega_0, \Gamma_0)$ as follows.

(1) For $\omega \in \Omega_1$ $G_0(f)(\omega) = \overline{\ }\sum\{f(\alpha) \mid \alpha \in g_1^{-1}(\omega)\}$
In some cases of aggregation it is important to offset the accounts of transactions. Then we introduce the offset function as follows.
(2) Let f be a transaction between ω_1 and ω_2 and AGS = {{<Deposits, MOU>, <Deposits payable, MOU>}, {<Interest earned, MOU>, <Interest expense, MOU>}}
Then offset function $S[e_1, e_2]$, $\{e_1, e_2\} \in$ AGS is defined as follows.

$$\text{For } f(\omega_1) = xe_1 + \ldots,\ f(\omega_2) = xe_2 + \ldots,\ \omega_1, \omega_2 \in g_1^{-1}(\omega),$$

$$S[e_1, e_2](f)(\omega_1) = \overline{\ }(f(\omega_1) + x\hat{\ }e_1)$$

$$S[e_1, e_2](f)(\omega_2) = \overline{\ }(f(\omega_2) + x\hat{\ }e_2)$$

We apply offset function at first and apply G_0 later. This is called consolidation in SNA. G_0 is called combination in SNA.

[Example 6.4.2] Simple-Sector-Aggregation Map G_0
As an example of $G_0: F(\Omega_0, \Gamma_0) \rightarrow F(\Omega_1, \Gamma_0)$ we adopt the case of Example 6.4.1.
Then $G_0(f[st, t_1])$(Industry) = 900<Cash, MOU> + 800<Machinery, MOU> + 950<Capital stock, MOU> + 750<Retained Earnings, MOU>.

6.5 National Economic Field and Bottom Up Reconstruction of SNA

6.5.1 Sector-National Economic Field: Flow Quantities

We introduce fsum as follows.

[Definition 6.5.1] fsum
fsum[e, SET] = $\sum$ {Project[<e, MOU>](f) | $f \in$ SET}, SET $\subseteq F[\Omega, \Gamma]$
Where Project[<e, MOU>](f) = $\sum$ {Project[<e, MOU>](f(ω)) | $\omega \in \Omega$} $\in [\Gamma]$.

Project[ω](f) = f(ω) ∈ [Γ],
Project[ω](SET) = ∑ {Project[ω](f) | f ∈ SET} for ω ∈ Ω.
As "SET" we introduce the following sets of transactions.

(1) Production Transaction Set
= {f[Production: ep(ω)](ω) | ω ∈ Industry}
Government Production Transaction Set
= {f[Government Service Production](Government)}

(2) Import Transaction Set
= {f[Import: em(ω)](ω) | ω ∈ {Flour-milling Industry, Bread Industry, Machinery Industry}} ∪ {f[Import: Machinery](ω)| ω ∈ Industry ∪ {Government, Household}} ∪ {f[Import: Bread](Household)}
Product Export Transaction Set
= {f[Export: ep(ω)](ω) | ω ∈ Industry}

(3) Value added Adjustment Transaction Set
= {f[Value added_Product Stock Adjustment](ω)| ω ∈ Industry}

(4) Stock Investment Set
= {f[Product Stock Investment Transfer: ep(ω)](ω) | ω ∈ Industry} ∪ {f[Material Stock Investment Transfer: em(ω)](ω) | ω ∈ {Flour-milling Industry, Bread Industry, Machinery Industry}}

(5) Household Consumption Set
= {f[Household_Consumption expenditure](Household)}
Government Consumption Expenditure Set
= {f[Government Service Expense](Government))}

(6) Equipment Investments Set
= {f[Equipment investments Transfer: Machinery](ω) | ω ∈ Industry ∪ {Government, Household}}
Industry Equipment Investments Set
= ∪ {f[Sell: Machinery: ω](Machinery Industry) | ω ∈ Industry}}
Domestic Equipment Investments Set
= {f[Sell: Machinery: ω](Machinery Industry) | ω ∈ Industry ∪ {Government, Household}}

(7) Wage Payment Transaction Set
= {f[Employment: ω](ω) | ω ∈ Industry ∪ {Government}}
Government Wage Payment Transaction Set
= {f[Employment: Government](Government)}
Industry Wage Payment Transaction Set
= {f[Employment: ω](ω) | ω ∈ Industry}
Household Wages Earned Transaction Set
= {f[Employment: ω](Household) | ω ∈ Industry ∪ {Government}}

(8) Depreciation Set
= {f[Depreciation: ω](ω) | ω ∈ Industry ∪ {Government, Household}}
Industry Depreciation Set = {f[Depreciation: ω](ω) | ω ∈ Industry}

Government Depreciation Set
= {f[Depreciation: Government](Government)}
Household Depreciation Set
= {f[Depreciation: Household](Household)}

(9) Transfer of Gross Profit Set
= {f[Transfer of Gross Profit: ω](ω) | ω ∈ Industry}

(10) Industry Interest Earned Set
= {f[Deposit Interest: ω](ω), f[National Bond Interest: ω](ω) | ω ∈ Industry}
Household Interest Earned Set
= {f[National Bond Interest: Household](Household), f[Deposit Interest: Household](Household)}
Bank Interest Earned Set
= {f[Loan Interest Expense: ω](Bank) | ω ∈ Industry ∪ {Household}}
Industry Interest Expense Set
= {f[Loan Interest Expense: ω](ω) | ω ∈ Industry}
Household Interest Expense Set
= {f[Loan Interest Expense: Household](Household)}
Government Interest Expense Set
= {f[National Bond Interest: ω](Government) | ω ∈ {Household} ∪ Industry}
Bank Interest Expense Set
= {f[Deposit Interest: ω](Bank) | ω ∈ Industry ∪ {Household}}
∪ {f[Central Bank Loan Interest Expense](Bank)}

(11) Industry Taxes Expense Set = {f[Tax: ω](ω) | ω ∈ Industry}
Household Taxes Expense Set = {f[Tax: Household](Household)}
Bank Taxes Expense Set = {f[Tax: Bank](Bank)}
Government Tax Revenue Set
= {f[Tax: ω](Government) | ω ∈ Industry ∪ {Household, Bank}}

(12) Industry Retained Earnings Set
= {f[Retained Earnings: ω](ω) | ω ∈ Industry}
Household Retained Earnings Set
= {f[Retained Earnings: Household](Household)}
Bank Retained Earnings Set
= {f[Retained Earnings: Bank}](Bank)}
Government Retained Earnings Set
= {f[Retained Earnings: Government}](Government)}

(13) Subsidy Income Set
= {f[Government Subsidy: ω](ω) | ω ∈ Industry ∪ {Household, Bank}}
Industry Subsidy Income Set
= {f[Government Subsidy: ω](ω) | ω ∈ Industry}
Household Subsidy Income Set

= {f[Government Subsidy: Household](Household)}
Bank Subsidy Income Set = {f[Government Subsidy: Bank](Bank)}
Government Subsidy Expenses Set
= {f[Government Subsidy: ω](Government) | ω ∈ Industry ∪ {Household, Bank}}

(14) Government Central Bank Payment Set
= {f[Central Bank Payment](Government)}

[Definition 6.5.2] Production Vector, Needs Vector
Let e ∈ Products = {$e_1, \ldots, e_k$}. Then the production Vector "fpro[e]" and the needs Vector "fneeds[e]" are defined as follows.
Where e = ep(β) = WhatProduct(β), e ∈ Products.
This means that "e" is produced by β. Let "e" be intermediate input then "e" is produced by β and is sold to α and becomes material in α.
Let e = ep(β) = em(α) = WhatMaterial(α).

(1) Production Vector fpro[e]
fpro[e]
= fsum[e, Production Transaction Set]
+ fsum[e, Import Transaction Set]
+ fsum[e, Stock Investment Set]

Where

(A)
fsum[e, Production Transaction Set]
= ∑ {Project[<e, MOU>](f) | f ∈ Production Transaction Set
= ∑ {Project[<e, MOU>](f) | f ∈ {f[Production: ep(ω)](ω) | ω ∈ Industry}}
= Project[<e, MOU>](f[Production: e](β))
= ProPrice[ep(β)](t)<ep(β), MOU>

(B-1) "e": Intermediate Consumption
fsum[e, Import Transaction Set]
= ∑ {Project[<e, MOU>](f) | f ∈ Import Transaction Set}
= Project[<em(α), MOU>](f[Import: em(α)](α))
= Import[em(α)](α, t)<em(α), MOU>
(B-2) "e": Equipment Investment(Machinery)
fsum[Machinery, Import Transaction Set]
= ∑ {Import[Machinery](ω, t)<Machinery, MOU> | ω ∈ Industry ∪ {Household, Government}}
(B-3) "e": Consumer Consumption (Bread)
fsum[Bread, Import Transaction Set]
= Import[Bread](Household, t)<Bread, MOU>

(C-1) "e": Intermediate Consumption
fsum[e, Stock Investment Set]
= ∑ {Project[<e, MOU>](f) | f ∈ Stock Investment Set}

= Project[<ep(β), MOU>](f[Product Stock Investment Transfer: ep(β)](β))
+ Project[<em(α), MOU>](f[Material Stock Investment Transfer: em(α)](α))
= {Max(0, –I[Product Stock Investment: ep(β)](β))
+ Max(0, –I[Material Stock Investment: em(α)](α))}<e, MOU>
(C-2) "e": Equipment Investment(Machinery)
fsum[e, Stock Investment Set]
= Max(0, –I[Product Stock Investment: Machinery](Machinery Industry))
<Machinery, MOU>
(C-3) "e": Consumer Consumption (Bread)
fsum[e, Stock Investment Set]
= Max(0, –I[Product Stock Investment: Bread](Bread Industry))
<Bread, MOU>

(2) Needs Vector fneeds[e]
(2-1) Let "e" be an intermediate consumption.
Then we assume that "e" is made by β and used by α as a material.
Thus e = ep(β) = em(α) and
fneeds[e]
= fsum[^em(α), Production Transaction Set]
+ fsum[^em(α), Stock Investment Set]
+ fsum[^ep(β), Product Export Transaction Set].
Where,
(A)
fsum[^em(α), Production Transaction Set]
= ∑{Project[^<em(α), MOU>](f) | f ∈ Production Transaction Set}
= Project[^<em(α), MOU>](f[Production: em(α)](α))
= Use[em(α)](t)^<em(α), MOU>
(B)
fsum[^e, Stock Investment Set]
= ∑{Project[^<e, MOU>](f) | f ∈ Stock Investment Set}
= {Project[^<ep(β), MOU>](f[Product Stock Investment Transfer: ep(β)](β))
+ Project[^<em(α), MOU>](f[Material Stock Investment Transfer: em(α)]
(α))}^<ep(β), MOU>
= {Max(0, I[Product Stock Investment: ep(β)](β))
+ Max(0, I[Material Stock Investment: em(α)](α))}^<e, MOU>
(C)
fsum[^ep(β), Product Export Transaction Set]
= Project[^<ep(β), MOU>](f[Export: ep(β)](ω))
= Export[ep(β)](β, t)^<ep(β), MOU>

(2-2) "e": Equipment Investment(Machinery)
fneeds[Machinery]
= fsum[^Machinery, Equipment Investments Set]
+ fsum[^Machinery, Stock Investment Set]
+ fsum[^Machinery, Product Export Transaction Set]

(A)
fsum[^Machinery, Equipment Investments Set]
= ∑ {Project[^<Machinery, MOU>](f[Equipment investments Transfer: Machinery](ω) | ω ∈ Industry ∪ {Government, Household}}
= ∑{InvPrice[ω](t) | ω ∈ Industry ∪ {Government, Household}}
+ ∑ {Import[Machinery](ω, t) | ω ∈ Industry ∪ {Government, Household}}
<^Machinery, MOU>
(B)
fsum[^Machinery, Stock Investment Set]
= Max(0, I[Product Stock Investment: Machinery](Machinery))
^<Machinery, MOU>
(C)
fsum[^Machinery, Product Export Transaction Set]
= Export[Machinery](Machinery Industry, t)^<Machinery, MOU>

(2-3) "e": Consumer Consumption (Bread)
fneeds[Bread]
= fsum[^Bread, Stock Investment Set]
+ fsum[^Bread, Product Export Transaction Set]
+ fsum[^Bread, Household Consumption Set]
(A)
fsum[^Bread, Stock Investment Set]
= ∑ {Project[<^Bread, MOU>](f) | f ∈ Stock Investment Set}
= Max(0, I[Product Stock Investment: Bread](Bread))^<Bread, MOU>
(B)
fsum[^Bread, Product Export Transaction Set]
= Project[^<Bread, MOU>](f[Export: Bread]((Bread Industry))
^<Bread, MOU>
= Export[Bread](Bread Industry, t)^<Bread, MOU> Export[Bread]
(Bread Industry, t)^<Bread, MOU>
(C)
fsum[^Bread, Household Consumption Set]
= ∑ {Project[^<Bread, MOU>](f) | f ∈ Household Consumption Set}
= {HouseDomCons(t) + Import[Bread](Household, t)}^<Bread, MOU>

Then
fneeds[e] = fsum[^e, Production Transaction Set]
+ fsum[^e, Equipment Investments Set]
+ fsum[^e, Stock Investment Set]
+ fsum[^e, Product Export Transaction Set]
+ fsum[^e, Household Consumption Set]

[Proposition 6.5.1]
|fpro[e]| = |fneeds[e]|

Proof:
We separate the proof into the following three cases.

(A)
Let "e" be used for intermediate consumption.
fsum[e, Production Transaction Set]
= ProPrice[ep(β)](t)<ep(β), MOU>
fsum[e, Import Transaction Set]
= Import[em(α)](α, t)<em(α), MOU>

|fpro[e]| = ProPrice[ep(β)](t)
+ Import[em(α)](α, t) + Max(0, –I[Product Stock Investment: ep(β)](β))
+ Max(0, –I[Material Stock Investment: em(α)](α))
|fneeds[e]|
= Use[em(α)](t) + Max(0, I[Product Stock Investment: ep(β)](β))
+ Max(0, I[Material Stock Investment: em(α)](α)) + Export[ep(β)](β, t)

Where,

I[Product Stock Investment: ep(β)](β)
= Use[em(β)](t) + ValueAdd[β](t)
– SellPrice[ep(β)](t) – Export[ep(β)](β, t)
I[Material Stock Investment: em(α)](α)
= SellPrice[em(α)](t)
+ Import[em(α)](α, t) – Use[em(α)](t)

Thus, I[Product Stock Investment: ep(β)](β) and I[Material Stock Investment: em(α)](α) must be classified depending on the positive or negative sign and we will demonstrate a positive case.

|fneeds[e]|
= Use[em(α)](t) + Export[ep(β)](β, t) + Use[em(β)](t)
+ ValueAdd[β](t) – SellPrice[ep(β)](t) – Export[ep(β)](β, t)
+ SellPrice[em(α)](t) + Import[em(α)](α, t) – Use[em(α)](t)
= Use[em(β)](t) + ValueAdd[β](t) + Import[em(α)](α, t)
= ProPrice[ep(β)](t) + Import[em(α)](α, t)
= |fpro[e]|
Where Use[em(β)](t) + ValueAdd[β](t) = ProPrice[ep(β)](t)

(B)
Let "e" denote equipment investment(Machinery)
fsum[e, Production Transaction Set]
= ProPrice[ep(ω)](t)<Machinery, MOU>
fsum[e, Import Transaction Set]
= ∑{Import[Machinery](ω, t)<Machinery, MOU> | ω ∈ Industry ∪ {Household, Government}}<Machinery, MOU>

|fpro[e]|
= |fsum[e, Production Transaction Set] + fsum[e, Import Transaction Set]
+ fsum[e, Stock Investment Set]|
= ProPrice[ep(ω)](t) + ∑ {Import[Machinery](ω, t)<Machinery, MOU>
| ω ∈ Industry ∪ {Household, Government}}
+ Max(0, –I[Product Stock Investment: Machinery](Machinery))

|fneeds[e]|
= |fsum[^Machinery, Equipment Investments Set]
+ fsum[^Machinery, Stock Investment Set]
+ fsum[^Machinery, Product Export Transaction Set]|
= ∑{InvPrice[ω](t) | ω ∈ Industry ∪ {Government, Household}}
+ ∑ {Import[Machinery](ω, t) | ω ∈ Industry ∪ {Government, Household}}
+ Max(0, I[Product Stock Investment: Machinery](Machinery))
+ Export[Machinery](Machinery Industry, t)

However, I[Product Stock Investment: Machinery](Machinery Industry)
= ProPrice[Machinery](t)
− ∑ {InvPrice[ω](t) | ω ∈ Industry ∪ {Household, Government}}
− Export[Machinery](Machinery Industry, t)

We have to distinguish two cases. Now we demonstrate that I[Product Stock Investment: Machinery](Machinery Industry) is a positive.

|fneeds[e]|
= ∑ {InvPrice[ω](t) | ω ∈ Industry ∪ {Government, Household}}
+ ∑ {Import[Machinery](ω, t) | ω ∈ Industry ∪ {Government, Household}}
+ Export[Machinery](Machinery Industry, t)
+ I[Product Stock Investment: Machinery](Machinery)
= ∑ {Import[Machinery](ω, t) | ω ∈ Industry ∪ {Government, Household}}
+ ProPrice[Machinery](t)
= |fpro[e]|

(C)
Let "e" denote consumption expenditure (Bread).
fsum[Bread, Production Transaction Set]
= ProPrice[Bread](t)<Bread, MOU>
fsum[Bread, Import Transaction Set]
= Import[Bread](Household, t)<Bread, MOU>

|fpro[Bread]|
= ProPrice[Bread](t) + Import[Bread](Household, t)
+ Max(0, −I[Product Stock Investment: ep(ω)](ω))
|fneeds[Bread]|
= Max(0, I[Product Stock Investment: Bread](Bread Industry))
+ Export[Bread](Bread Industry, t) + HouseDomCons(t)
+ Import[Bread](Household, t)

Where I[Product Stock Investment: Bread](ω) = ProPrice[Bread](t)
− SellPrice[Bread](t) − Export[Bread](Bread Industry, t)
Then we ascertain whether I[Product Stock Investment: Bread](Bread Industry) is a negative or positive number. We will demonstrate the case that it is positive. From HouseDomCons(t) = SellPrice[Bread](t),
|fneeds[Bread]|
= ProPrice[Bread](t) − SellPrice[Bread](t)
− Export[Bread](Bread Industry, t) + Export[Bread](Bread Industry, t)

+ HouseDomCons(t) + Import[Bread](Household, t)
= ProPrice[Bread](t) + Import[Bread](Household, t)
= |fpro[Bread]|. Q.E.D.

Let fIntermediate Input[e] = fsum [^e, Product Export Transaction Set]. Because fsum[^e, Product Export Transaction Set] denotes intermediate consumption.

TABLE 6.5.1 |fpro[e]| = |fneeds[e]|

fsum[e, Production Transaction Set] fsum[e, Import Transaction Set] fsum[e, Stock Investment Set]	fIntermediate Input[e] fsum[^e, Equipment Investments Set] fsum[^e, Government Equipment Investments Set] fsum[^e, Stock Investment Set] fsum[^e, Product Export Transaction Set] fsum[^e, Household Consumption Set]
fpro[e]	fneeds[e]

[Example 6.5.1] Example of a Small National Economic Field
[Production]
f[Production: Wheat](Agriculture)
= 100<Wheat, MOU> + 100<Value added, MOU>
f[Production: Wheat flour](Flour-milling Industry)
= 100^<Wheat, MOU> + 200<Wheat flour, MOU> + 100<Value added, MOU>
f[Production: Bread](Bread Industry)
= 200^<Wheat flour, MOU> + 300<Bread, MOU> + 100<Value added, MOU>
f[Production: Steel](Steel Industry)
= 40<Steel, MOU> + 40<Value added, MOU>
f[Production: Machinery](Machinery Industry)
= 50^<Steel, MOU> + 160<Machinery, MOU> + 110<Value added, MOU>
f[Production: Government Service](Government)
= 50<Government Service, MOU> + 50<Value added, MOU>

[Employment]
f[Employment: Agriculture](Agriculture)
= 80<Wages expenditure, MOU> + 80^<Cash, MOU>
f[Employment: Agriculture](Household)
= 80<Wages earned, MOU> + 80<Cash, MOU>
f[Employment: Flour-milling Industry](Flour-milling Industry)
= 80<Wages expenditure, MOU> + 80^<Cash, MOU>
f[Employment: Flour-milling Industry](Household)
= 80<Wages earned, MOU> + 80<Cash, MOU>
f[Employment: Bread Industry](Bread Industry)
= 80<Wages expenditure, MOU> + 80^<Cash, MOU>

f[Employment: Bread Industry](Household)
= 80<Wages earned, MOU> + 80<Cash, MOU>
f[Employment: Steel Industry](Steel Industry)
= 20<Wages expenditure, MOU> + 20^<Cash, MOU>
f[Employment: Steel Industry](Household)
= 20<Wages earned, MOU> + 20<Cash, MOU>
f[Employment: Machinery Industry](Machinery Industry)
= 90<Wages expenditure, MOU> + 90^<Cash, MOU>
f[Employment: Machinery Industry](Household)
= 90<Wages earned, MOU> + 90<Cash, MOU>
f[Employment: Government](Government)
= 50<Wages expenditure, MOU> + 50^<Cash, MOU>
f[Employment: Government](Household)
= 50<Wages earned, MOU> + 50<Cash, MOU>

[Tax Payment]
f[Tax Transaction: Agriculture](Agriculture)
= 2<Taxes expense, MOU> + 2^<Cash, MOU>
f[Tax Transaction: Agriculture](Government)
= 2<Taxes revenue, MOU> + 2<Cash, MOU>
f[Tax Transaction: Flour-milling Industry](Flour-milling Industry)
= 2<Taxes expense, MOU> + 2^<Cash, MOU>
f[Tax Transaction: Flour-milling Industry](Government)
= 2<Taxes revenue, MOU> + 2<Cash, MOU>
f[Tax Transaction: Bread Industry](Bread Industry)
= 2<Taxes expense, MOU> + 2^<Cash, MOU>
f[Tax Transaction: Bread Industry](Government)
= 2<Taxes revenue, MOU> + 2<Cash, MOU>
f[Tax Transaction: Steel Industry](Steel Industry)
= 2<Taxes expense, MOU> + 2^<Cash, MOU>
f[Tax Transaction: Steel Industry](Government)
= 2<Taxes revenue, MOU> + 2<Cash, MOU>
f[Tax Transaction: Machinery Industry(Machinery Industry)
= 2<Taxes expense, MOU> + 2^<Cash, MOU>
f[Tax Transaction: Machinery Industry](Government)
= 2<Taxes revenue, MOU> + 2<Cash, MOU>
f[Tax Transaction: Household](Household)
= 40<Taxes expense, MOU> + 40^<Cash, MOU>
f[Tax Transaction: Household](Government)
= 40<Taxes revenue, MOU> + 40<Cash, MOU>
f[Tax Transaction: Bank](Bank)
= 2<Taxes expense, MOU> + 2^<Cash, MOU>
f[Tax Transaction: Bank](Government)
= 2<Taxes revenue, MOU> + 2<Cash, MOU>

[Bank Loans]
f[Loan: Agriculture](Bank)
= 2<Loans receivable, MOU> + 2^<Cash, MOU>
f[Loan: Agriculture](Agriculture)
= 2<Loans payable, MOU> + 2<Cash, MOU>
f[Loan: Flour-milling Industry](Bank)
= 12<Loans receivable, MOU> + 12^<Cash, MOU>
f[Loan: Flour-milling Industry](Flour-milling Industry)
= 12<Loans payable, MOU> + 12<Cash, MOU>
f[Loan: Bread Industry](Bank)
= 12<Loans receivable, MOU> + 12^<Cash, MOU>
f[Loan: Bread Industry](Agriculture)
= 12<Loans payable, MOU> + 12<Cash, MOU>
f[Loan: Steel Industry](Bank)
= 12<Loans receivable, MOU> + 12^<Cash, MOU>
f[Loan: Steel Industry](Steel Industry)
= 12<Loans payable, MOU> + 12<Cash, MOU>
f[Loan: Machinery Industry](Bank)
= 2<Loans receivable, MOU> + 2^<Cash, MOU>
f[Loan: Machinery Industry](Machinery Industry)
= 2<Loans payable, MOU> + 2<Cash, MOU>

[Deposits]
f[Deposit: Agriculture](Agriculture)
= 18^<Cash, MOU> + 18<Deposits, MOU>
f[Deposit: Agriculture](Bank)
= 18<Cash, MOU> + 18<Deposits payable, MOU>
f[Deposit: Flour-milling Industry](Flour-milling Industry)
= 18^<Cash, MOU> + 18<Deposits, MOU>
f[Deposit: Flour-milling Industry](Bank)
= 18<Cash, MOU> + 18<Deposits payable, MOU>
f[Deposit: Bread Industry](Bread Industry)
= 18^<Cash, MOU> + 18<Deposits, MOU>
f[Deposit: Bread Industry](Bank)
= 18<Cash, MOU> + 18<Deposits payable, MOU>
f[Deposit: Steel Industry](Steel Industry)
= 18^<Cash, MOU> + 18<Deposits, MOU>
f[Deposit: Steel Industry](Bank)
= 18<Cash, MOU> + 18<Deposits payable, MOU>
f[Deposit: Machinery Industry](Machinery Industry)
= 18^<Cash, MOU> + 18<Deposits, MOU>
f[Deposit: Machinery Industry](Bank)
= 18<Cash, MOU> + 18<Deposits payable, MOU>
f[Deposit: Household](Household)
= 50^<Cash, MOU> + 50<Deposits, MOU>

f[Deposit: Household](Bank)
= 50<Cash, MOU> + 50<Deposits payable, MOU>

[National Bond Purchase]

f[National Bond Purchase: Household](Household)
= 10<National Bonds, MOU> + 10^<Cash, MOU>
f[National Bond Purchase: Household](Government)
= 10<National Bonds payable, MOU> + 10<Cash, MOU>

[Equipment Investments]

f[Domestic Equipment investments: Machinery: Agriculture](Agriculture)
= 20^<Cash, MOU> + 20<Machinery, MOU>
f[Sell: Machinery: Agriculture](Machinery Industry)
= 20<Cash, MOU> + 20^<Machinery, MOU>
f[Domestic Equipment investments: Machinery: Flour-milling Industry]
(Flour-milling Industry)
= 30^<Cash, MOU> + 30<Machinery, MOU>
f[Sell: Machinery: Flour-milling Industry](Machinery Industry)
= 30<Cash, MOU> + 30^<Machinery, MOU>
f[Domestic Equipment investments: Machinery: Bread Industry]
(Bread Industry)
= 30^<Cash, MOU> + 30<Machinery, MOU>
f[Sell: Machinery: Bread Industry](Machinery Industry)
= 30<Cash, MOU> + 30^<Machinery, MOU>
f[Domestic Equipment investments: Machinery: Steel Industry](Steel Industry)
= 30^<Cash, MOU> + 30<Machinery, MOU>
f[Sell: Machinery: Steel Industry](Machinery Industry)
= 30<Cash, MOU> + 30^<Machinery, MOU>
f[Domestic Equipment investments: Machinery: Machinery Industry]
(Machinery Industry)
= 30^<Cash, MOU> + 30<Machinery, MOU>
f[Sell: Machinery: Machinery Industry](Machinery Industry)
= 30<Cash, MOU> + 30^<Machinery, MOU>
f[Domestic Equipment investments: Machinery: Government](Government)
= 10^<Cash, MOU> + 10<Machinery, MOU>
f[Sell: Machinery: Government](Machinery Industry)
= 10<Cash, MOU> + 10^<Machinery, MOU>

[Domestic Product Transactions]

f[Sell: Wheat: Flour-milling Industry](Agriculture)
= 100^<Wheat, MOU> + 100<Cash, MOU>
f[Buy: Wheat: Agriculture](Flour-milling Industry)
= 100<Wheat, MOU> + 100^<Cash, MOU>
f[Sell: Wheat flour: Bread Industry](Flour-milling Industry)
= 200^<Wheat flour, MOU> + 200<Cash, MOU>

f[Buy: Wheat flour: Flour-milling Industry](Bread Industry)
= 200<Wheat flour, MOU> + 200^<Cash, MOU>
f[Sell: Bread: Household](Bread Industry)
= 300^<Bread, MOU> + 300<Cash, MOU>
f[Buy: Bread: Bread Industry](Household)
= 300<Bread, MOU> + 300^<Cash, MOU>
f[Sell: Steel: Machinery Industry](Steel Industry)
= 40^<Steel, MOU> + 40<Cash, MOU>
f[Buy: Steel: Steel Industry](Machinery Industry)
= 40<Steel, MOU> + 40^<Cash, MOU>

[Export & Import Transactions]
f[Import: Steel](Machinery Industry)
= 10<Steel, MOU> + 10^<Cash, MOU>
f[Import: Steel](Oversea)
= 10^<Steel, MOU> + 10<Cash, MOU>
f[Export: Machinery](Machinery Industry)
= 10^<Machinery, MOU> + 10<Cash, MOU>
f[Export: Machinery](Oversea)
= 10<Machinery, MOU> + 10^<Cash, MOU>

[Consumption Expenditure]
f[Household Consumption expenditure](Household)
= 300^<Bread, MOU> + 300<Consumption expenditure, MOU>
f[Government Service Expense](Government)
= 50^<Government Service, MOU> + 50<Consumption expenditure, MOU>

[Example 6.5.2] Production and Needs Vector
(1) Steel: Production and Needs fpro[Steel]
fsum[Steel, Production Transaction Set] = 40<Steel, MOU>
fsum[Steel, Import Transaction Set] = 10<Steel, MOU>
fpro[Steel]
= fsum[Steel, Production Transaction Set]
+ fsum[Steel, Import Transaction Set] = 50<Steel, MOU>
fIntermediate Input[Steel] = 50^<Steel, MOU>
fsum[^Steel, Industry Equipment Investments Set] = 0
fsum[^Steel, Government Equipment Investments Set] = 0
fsum[^Steel, Stock Investment Set] = 0
fsum[^Steel, Product Export Transaction Set] = 0
fsum[^Steel, Household Consumption Set] = 0
fneeds[Steel] = 50^<Steel, MOU>
|fpro[Steel]| = |fneeds[Steel]| = 50

(2) Machinery: Production and Needs
fsum[Machinery, Production Transaction Set] = 160<Machinery, MOU>

fsum[Machinery, Import Transaction Set] = 0
fpro[Machinery] = 160<Machinery, MOU>
fIntermediate Input[Machinery] = 0
fsum[^Machinery, Industry Equipment Investments Set]
= 30^<Machinery, MOU> + 30^<Machinery, MOU>
+ 30^<Machinery, MOU> + 30^<Machinery, MOU>
+ 20^<Machinery, MOU> = 140^<Machinery, MOU>
fsum[^Machinery, Government Equipment Investments Set]
= 10^<Machinery, MOU>
fsum[^Machinery, Stock Investment Set] = 0
fsum[^Machinery, Product Export Transaction Set] = 10^<Machinery, MOU>
fsum[^Machinery, Household Consumption Set] = 0
fneeds[Machinery] = 160^<Machinery, MOU>
|fpro[Machinery]| = |fneeds[Machinery]| = 160

(3) Wheat: Production and Needs
fsum[Wheat, Production Transaction Set] = 100<Wheat, MOU>
fsum[Wheat, Import Transaction Set] = 0
fpro[Wheat] = 100<Wheat, MOU>
fIntermediate Input[Wheat] = 100^<Wheat, MOU>
fsum[^Wheat, Industry Equipment Investments Set] = 0
fsum[^Wheat, Government Equipment Investments Set] = 0
fsum[^Wheat, Stock Investment Set] = 0
fsum[^Wheat, Product Export Transaction Set] = 0
fsum[^Wheat, Household Consumption Set] = 0
fneeds[Wheat] = 100^<Wheat, MOU>
|fpro[Wheat]| = |fneeds[Wheat]| = 100

(4) Wheat flour: Production and Needs
fsum[Wheat flour, Production Transaction Set] = 200^<Wheat flour, MOU>
fsum[Wheat flour, Import Transaction Set] = 0
fpro[Wheat flour] = 200<Wheat flour, MOU>
fIntermediate Input[Wheat flour] = 200^<Wheat flour, MOU>
fsum[^Wheat flour, Industry Equipment Investments Set] = 0
fsum[^Wheat flour, Government Equipment Investments Set] = 0
fsum[^Wheat flour, Stock Investment Set] = 0
fsum[^Wheat flour, Product Export Transaction Set] = 0
fsum[^Wheat flour, Household Consumption Set] = 0
fneeds[Wheat flour] = 200^<Wheat flour, MOU>
|fpro[Wheat flour]| = |fneeds[Wheat flour]| = 200

(5) Bread: Production and Needs
fsum[Bread, Production Transaction Set] = 300<Bread, MOU>
fsum[Bread, Import Transaction Set] = 0
fpro[Bread] = 300<Bread, MOU>
fIntermediate Input[Bread] = 0

fsum[^Bread, Industry Equipment Investments Set] = 0
fsum[^Bread, Government Equipment Investments Set] = 0
fsum[^Bread, Stock Investment Set] = 0
fsum[^Bread, Product Export Transaction Set] = 0
fsum[^Bread, Household Consumption Set] = 300^<Bread, MOU>
fneeds[Bread] = 300^<Bread, MOU>
|fpro[Bread]| = |fneeds[Bread]| = 300

We will use the same data of this example later on.

[Definition 6.5.3] Product Integrated Production Vector, Product Integrated Needs Vector

Product Integrated Production Vector fpro[Product] and Product Integrated Needs Vector fneeds[Product] are defined as follows.

(1) fpro[Product] = Σ{fpro[e] | e ∈ Products}
(2) fpro[Domestic Product]
= Σ{fsum[e, Production Transaction Set] | e ∈ Products}
(3) fpro[Overseas Product]
= Σ{fsum[e, Import Transaction Set] | e ∈ Products}
(4) fneeds[Product]
= fneeds[Production Transaction Set]
+ fneeds[Industry Equipment Investments Set]
+ fneeds[Government Equipment Investments Set]
+ fneeds[Stock Investment Set]
+ fneeds[Product Export Transaction Set]
+ fneeds[Household Consumption Set]
(5) fneeds[Production Transaction Set]
= Σ{fsum[^e, Production Transaction Set] | e ∈ Products}
(6) fneeds[Industry Equipment Investments Set]
= Σ{fsum[^e, Industry Equipment Investments Set] | e ∈ Products}
(7) fneeds[Government Equipment Investments Set]
= Σ{fsum[^e, Government Equipment Investments Set] | e ∈ Products}
(8) fneeds[Stock Investment Set]
= Σ{fsum[^e, Stock Investment Set] | e ∈ Products}
(9) fneeds[Product Export Transaction Set]
= Σ{fsum[^e, Product Export Transaction Set] | e ∈ Products}
(10) fneeds[Household Consumption Set]
= Σ{fsum[^e, Household Consumption Set] | e ∈ Product}

[Cor.]
fpro[Product] = fpro[Domestic Product] + fpro[Overseas Product]

Proof:
It is clear from fpro[e] = fsum[e, Production Transaction Set]
+ fsum[e, Export & Import Transaction]. Q.E.D.

[Example 6.5.3] Product Integrated Needs Vector, Product Integrated Production Vector, Product Integrated Needs Vector

(1) fpro[Products]

fpro[Domestic Product]
= fsum[Steel, Production Transaction Set]
+ fsum[Machinery, Production Transaction Set]
+ fsum[Wheat, Production Transaction Set]
+ fsum[Wheat flour, Production Transaction Set]
+ fsum[Bread, Production Transaction Set]
= 40<Steel, MOU> + 160<Machinery, MOU> + 100<Wheat, MOU>
+ 200<Wheat flour, MOU> + 300<Bread, MOU>

fpro[Overseas Product]
= fsum[Steel, Import Transaction Set]
+ fsum[Machinery, Import Transaction Set]
+ fsum[Wheat, Import Transaction Set]
+ fsum[Wheat flour, Import Transaction Set]
+ fsum[Bread, Import Transaction Set]
= 10<Steel, MOU> + 0<Machinery, MOU> + 0<Wheat, MOU>
+ 0<Wheat flour, MOU> + 0<Bread, MOU>

fpro[Products]
= fpro[Domestic Product] + fpro[Overseas Product]
= 50<Steel, MOU> + 160<Machinery, MOU> + 100<Wheat, MOU>
+ 200<Wheat flour, MOU> + 300<Bread, MOU>

(2) fneeds[Products]

fneeds[Production Transaction Set]
= fsum[^Steel, Production Transaction Set]
+ fsum[^Machinery, Production Transaction Set]
+ fsum[^Wheat, Production Transaction Set]
+ fsum[^Wheat flour, Production Transaction Set]
+ fsum[^Bread, Production Transaction Set]
= 50^<Steel, MOU> + 0^<Machinery, MOU> + 100^<Wheat, MOU>
+ 200^<Wheat flour, MOU> + 0^<Bread, MOU>

fneeds[Industry Equipment Investments Set] = 140^<Machinery, MOU>,
fneeds[Government Equipment Investments Set] = 10^<Machinery, MOU>,
fneeds[Stock Investment Set] = 0,
fneeds[Product Export Transaction Set]
= 10^<Machinery, MOU>,
fneeds[Household Consumption Set]
= 300^<Bread, MOU>

fneeds[Products]
= 50^<Steel, MOU> + 0^<Machinery, MOU>
+ 100^<Wheat, MOU> + 200^<Wheat flour, MOU> + 0^<Bread, MOU>

+ 140^<Machinery, MOU> + 10^<Machinery, MOU>
+ 10^<Machinery, MOU> + 300^<Bread, MOU>
= 50^<Steel, MOU>
+ 160^<Machinery, MOU> + 100^<Wheat, MOU>
+ 200^<Wheat flour, MOU> + 300^<Bread, MOU>
|fpro[Products]| = |fneeds[Products]| = 810

[Definition 6.5.4] V Vector Field, U Vector Field
(1) V Vector Field f_V
V Vector Field shows which products are produced by which industry.
$f_V(\omega)$ denotes the production at ω ∈ Industry.
$f_V(\omega)$ = Σ{fsum[e, Project[ω](Production Transaction Set)] | e ∈ Products}
= Σ{fsum[e, {f(ω) | f ∈ Production Transaction Set}] | e ∈ Products}
= Σ{Σ{Project[<e, MOU>](f(ω)) | f ∈ Production Transaction Set} | e ∈ Products}

(2) U Vector Field f_U
$f_U(\omega)$ denotes the input of material at ω ∈ Industry.
$f_U(\omega)$ = Σ{fsum[^e, Project[ω](Production Transaction Set)] | e ∈ Products}
= Σ{fsum[^e, {f(ω) | f ∈ Production Transaction Set}] | e ∈ Products}
= Σ{Σ{Project[<^e, MOU>](f(ω)) | f ∈ Production Transaction Set} | e ∈ Products}
Where Project[ω](Production Transaction Set)
= {Project[ω](f) | f ∈ Production Transaction Set}
= {f(ω) | f ∈ Production Transaction Set}

[Proposition 6.5.2]
Let each industry makes a single product then

$f_V(\omega)$ = fsum[ep(ω), Production Transaction Set] holds,

where the product is denoted by ep(ω).

Proof:
We assume that the product of ω is ep(ω). Then
$f_V(\omega)$ = Σ{fsum[e, Project[ω](Production Transaction Set)] | e ∈ Products}
= Σ{fsum[e, {f(ω) | f ∈ Production Transaction Set}] | e ∈ Products}
= fsum[ep(ω), Production Transaction Set] Q.E.D.

[Proposition 6.5.3]
Let each industry uses a single material then

$f_U(\omega)$ = fsum[^e, Production Transaction Set] holds,

where the material is denoted by em(ω).

Proof:
We assume that the material of the product of ω is e. Then
$f_U(\omega)$ = Σ{fsum[^e, Project[ω](Production Transaction Set)] | e ∈ Products}

= $\sum$\{fsum[^e, \{f(ω) | f $\in$ Production Transaction Set\}] | e $\in$ Products\}
= fsum[^e, Production Transaction Set] Q.E.D.

[Example 6.5.4]
(1) V Vector Field
Let Industry = \{Steel Industry, Machinery Industry, Agriculture, Flour-milling Industry, Bread Industry\}.
$f_V(\omega)$ = fsum[e, Production Transaction Set] holds in virtual economy.
f_V(Steel Industry)
= fsum[Steel, Production Transaction Set]
= 40<Steel, MOU>
f_V(Machinery Industry)
= fsum[Machinery, Production Transaction Set]
= 160<Machinery, MOU>
f_V(Agriculture)
= fsum[Wheat, Production Transaction Set]
= 100<Wheat, MOU>
f_V(Flour-milling Industry)
= fsum[Wheat flour, Production Transaction Set]
= 200<Wheat flour, MOU>
f_V(Bread Industry)
= fsum[Bread, Production Transaction Set]
= 300<Bread, MOU>

(2) U Vector Field
f_U denotes the input of materials for production, where Products = \{<Steel, MOU>, <Machinery, MOU>, <Wheat flour, MOU>, <Wheat, MOU>, <Bread, MOU>\}.
$f_U(\omega)$ denotes the input of materials for production at $\omega \in$ Industry. For example,
f_U(Machinery Industry)
= fsum[^Steel, Production Transaction Set]
= 50^<Steel, MOU>,
f_U(Flour-milling Industry)
= fsum[^Wheat, Production Transaction Set]
= 100^<Wheat, MOU>,
f_U(Bread Industry)
= fsum[^Wheat flour, Production Transaction Set]
= 200^<Wheat flour, MOU>.

These vectors are shown as matrix form, which are called U Table and V Table respectively in SNA as follows.

[Definition 6.5.5] V Table, U Table
V Table: V(ω, e) = |Project[e]($f_V(\omega)$)|
U Table: U(e, ω) = |Project[e]($f_U(\omega)$)|

[Example 6.5.5] V Table, U Table
(1) V Table
V(ω, e) = |Project [e](f_V(ω))|
V(Steel Industry, Steel) = 40<Steel, MOU>
V(Machinery Industry, Machinery) = 160<Machinery, MOU>
V(Agriculture, Wheat) = 100<Wheat, MOU>
V(Flour-milling Industry, Wheat flour) = 200<Wheat flour, MOU>
V(Bread Industry, Bread) = 300<Bread, MOU>

(2) U Table
U(e, ω) = |Project[e](f_U(ω))|
U(Steel, Machinery Industry) = 50^<Steel, MOU>
U(Wheat, Flour-milling Industry) = 100^<Wheat, MOU>
U(Wheat flour, Bread Industry) = 200^<Wheat flour, MOU>

[Definition 6.5.6] Product × Product Table
Input from Product e_i to e_j: Product × Product Table T(e_i, e_j) is defined from U Vector Field and V Vector Field as follows.

$$T(e_i, e_j) = \mathrm{Project}[e_i](f_U(\omega))$$

Where we assume that each industry makes single product from single material.

[Definition 6.5.7] Industry-Sector, Government-Sector Value Added Production and Appropriation Vector
(1a)
fValue added production[Industry]
= ‾{fsum[Value added, Production Transaction Set]
+ fsum[^Value added, Value added Adjustment Transaction Set]
+ fsum[Value added, Value added Adjustment Transaction Set]}
Notice: ^<Value added, MOU> is used when the price of product has fallen.

(1b)
fValue added appropriation[Industry]
= fsum[Wages expenditure, Industry Wage Payment Transaction Set]
+ fsum[Reserve for depreciation, Industry Depreciation Set]
+ ‾{fsum[Gross Profit, Transfer of Gross Profit Set]
+ fsum[^Gross Profit, Transfer of Gross Profit Set]}

(2a)
fValue added production[ω], ω ∈ Industry
= ‾{fsum[Value added, {f[Production: ep(ω)](ω)}]
+ fsum[^Value added, {f[Value added_Product Stock Adjustment](ω)}]
+ fsum[Value added, {f[Value added_Product Stock Adjustment](ω)}]}
fValue added production[Industry] = ∑{fValue added production[ω]
| ω ∈ Industry}

(2b)
fValue added appropriation[ω], ω ∈ Industry
= fsum[Wages expenditure, {f[Employment: ω](ω)}]
+ fsum[Reserve for depreciation, {f[Depreciation: ω](ω)}]
+ ¬{fsum[Gross Profit, {f[Transfer of Gross Profit: ω](ω)}]
+ fsum[^Gross Profit, {f[Transfer of Gross Profit: ω](ω)}]}

(3a)
fValue added production[Government]
= fsum[Value added, Government Production Transaction Set]
= fsum[Value added, {f[Government Service Production](Government)}]

(3b)
fValue added appropriation[Government]
= fsum[Wages expenditure, {f[Employment: Government](Government)}]

[Example 6.5.6] Industry, Government: Value added production and appropriation Vector
fValue added production[ω] and fValue added appropriation[ω] are given in our example as follows.

fValue added production[Agriculture]
= 100<Value added, MOU>
fValue added production[Flour-milling Industry]
= 100<Value added, MOU>
fValue added production[Bread Industry]
= 100<Value added, MOU>
fValue added production[Steel Industry]
= 40<Value added, MOU>
fValue added production[Machinery Industry]
= 110<Value added, MOU>
fValue added production[Industry]
= 450<Value added, MOU>
fValue added production[Government]
= 50<Value added, MOU>
fValue added appropriation[Agriculture]
= 80<Wages expenditure, MOU> + 19<Gross Profit, MOU>
+ 1<Reserve for depreciation, MOU>
fValue added appropriation[Flour-milling Industry]
= 80<Wages expenditure, MOU> + 19<Gross Profit, MOU>
+ 1<Reserve for depreciation, MOU>
fValue added appropriation[Bread Industry]
= 80<Wages expenditure, MOU> + 19<Gross Profit, MOU>
+ 1<Reserve for depreciation, MOU>
fValue added appropriation[Steel Industry]
= 20<Wages expenditure, MOU> + 19<Gross Profit, MOU>
+ 1<Reserve for depreciation, MOU>

fValue added appropriation[Machinery Industry]
= 90<Wage, MOU> + 19<Gross Profit, MOU>
+ 1<Reserve for depreciation, MOU>
fValue added appropriation[Industry]
= 350<Wages expenditure, MOU> + 95<Gross Profit, MOU>
+ 5<Reserve for depreciation, MOU>
fValue added appropriation[Government]
= 50<Wages expenditure, MOU>

[Proposition 6.5.4]
|fValue added production[ω]| = |fValue added appropriation[ω]|
Where ω ∈ Industry or ω = Government.

Proof:
For ω ∈ Industry
|fValue added production[ω]|
= |‾{fsum[Value added, {f[Production: ep(ω)](ω)}]
+ fsum[^Value added, {f[Value added_Product Stock Adjustment](ω)}]
+ fsum[Value added, {f[Value added_Product Stock Adjustment](ω)}]}|
= |ValueAdd[ω](t) + ValAdjustUp[ω](t) − ValAdjustDown[ω](t)|

|fValue added appropriation[ω]|
= |fsum[Wages expenditure, {f[Employment: ω](ω)}]|
+ |fsum[Reserve for depreciation, {f[Depreciation: ω](ω)}]|
+ |‾{fsum[Gross Profit, {f[Transfer of Gross Profit: ω](ω)}]
+ fsum[^Gross Profit, {f[Transfer of Gross Profit: ω](ω)}]}|
= |Wages[ω](t) + Dep[ω](t) + {ValueAdd[ω] + ValAdjustUp[ω](t)
− Wages[ω](t) − Dep[ω](t) − ValAdjustDown[ω](t)}|
= |ValueAdd[ω](t) + ValAdjustUp[ω](t) − ValAdjustDown[ω](t)|
= |fValue added production[ω]|

Then |fValue added production[ω]| = |fValue added appropriation[ω]| holds.
Q.E.D.

For Government,
fsum[Value added, {f[Government Service Production](Government)}]
= Wages[Government](t) holds from the conventional definition.

[Proposition 6.5.5]
fValue added production[ω]
= Max(0, ($|f_V(\omega)|$ − $|f_U(\omega)|$ + I[Value added_Product Stock Adjustment](ω)))
<Value added, MOU>
+ Max(0, −($|f_V(\omega)|$ − $|f_U(\omega)|$ + I[Value added_Product Stock Adjustment](ω)))
^<Value added, MOU>

Proof:
We show such case that
$|f_V(\omega)|$ − $|f_U(\omega)|$ + I[Value added_Product Stock Adjustment](ω))≧ 0.

(1) $|f_V(\omega)|$ = |Σ{Σ{Project[<e, MOU>](f(ω)) | f ∈ Production Transaction Set} | e ∈ Products}|
= |Project[<ep(ω), MOU>](f[Production: ep(ω)](ω))|
= ProPrice[ep(ω)](t)

(2) $|f_U(\omega)|$ = |Σ{Σ{Project[<^e, MOU>](f(ω)) | f ∈ Production Transaction Set}| e ∈ Products}|
= |Project[<em(ω), MOU>](f[Production: ep(ω)](ω))|
= Use[em(ω)](t)

Where, ProPrice[ep(ω)](t) = Use[em(ω)](t) + ValueAdd[ω](t).
$|f_V(\omega)| - |f_U(\omega)|$ = ValueAdd[ω](t)
$|f_V(\omega)| - |f_U(\omega)|$ + I[Value added_Product Stock Adjustment](ω)
= ValueAdd[ω](t) + ValadjustUp[ω](t) – ValAdjustDown[ω](t)
|fValue added production[ω]|
= |ValueAdd[ω](t) + ValAdjustUp[ω](t) – ValAdjustDown[ω](t)|
fValue added production[ω] = ($|f_V(\omega)| - |f_U(\omega)|$
+ I[Value added_Product Stock Adjustment](ω))<Value added, MOU>

It is easy to show another case. Q.E.D.

[Example 6.5.7] Value Added by U Vector Field and V Vector Field
fValue added production[Agriculture]
= $|f_V$(Agriculture)| <Value added, MOU>
= 100<Value added, MOU>

fValue added production[Flour-milling Industry]
= ($|f_V$(Flour-milling Industry)| – $|f_U$(Flour-milling Industry)|)
<Value added, MOU>
= (|200<Wheat flour, MOU>| – |100^<Wheat, MOU>|)<Value added, MOU>
= 100<Value added, MOU>

fValue added production[Bread Industry]
= ($|f_V$(Bread Industry)| – $|f_U$(Bread Industry)|)<Value added, MOU>
= (|300<Bread, MOU>| – |200^<Wheat flour, MOU>|)<Value added, MOU>
= 100<Value added, MOU>

fValue added production[Steel Industry]
= $|f_V$(Steel Industry)| <Value added, MOU>
= 40<Value added, MOU>

fValue added production[Machinery Industry]
= ($|f_V$(Machinery Industry)| – $|f_U$(Machinery Industry)|)<Value added, MOU>
= (|160<Machinery, MOU>| – |50^<Steel, MOU>|)<Value added, MOU>
= 110<Value added, MOU>

[Definition 6.5.8] Production Vector, Cost Vector, Needs Vector
[1] Industry, Government Sector Production Vector fpro[ω]
[2] Industry, Government Sector Cost Vector fcost[ω]
[3] Industry, Government Sector Needs Vector fneeds[ω]

These are vectors of production and its cost, which are defined as follows.

(1a) fpro[Industry] = fpro[Domestic Product]
(1b) fcost[Industry] = $\sum\{f_U(\omega) \mid \omega \in \{a_1, \ldots, a_n\}\}$
+ fsum[Wages expenditure, Industry Wage Payment Transaction Set]
+ ¯{fsum[Gross Profit, Transfer of Gross Profit Set]
+ fsum[^Gross Profit, Transfer of Gross Profit Set]}
+ fsum[Depreciation, Industry Depreciation Set]

Where $\sum\{f_U(\omega) \mid \omega \in \{a_1, \ldots, a_n\}\}$ is called fInterim input. Then

fcost[Industry]
= fInterim input
+ fsum[Wages expenditure, Industry wages transaction]
+ fsum[Operating surplus, Industry Operating surplus transfer]
+ fsum[Reserve for depreciation, Industry Depreciation]

(1c) fneeds[Industry] = fneeds[Product]
(2a) fpro[Government]
= fsum[Government service, Government Production Transaction Set]
(2b) fcost[Government]
= fsum[Wages expenditure, Government Wage Payment Transaction Set]
+ fsum[Depreciation, Government Depreciation Set]
(2c) fneeds[Government]
= fsum[^Government service, GovernmentConsumption expenditure Set]

[Example 6.5.8] Example of Production Vector, Cost Vector, Needs Vector
(1a) fpro[Industry] = fpro[Domestic Product]
= 40<Steel, MOU> + 160<Machinery, MOU> + 100<Wheat, MOU>
+ 200<Wheat flour, MOU> + 300<Bread, MOU>
(1b) fcost[Industry] = 50^<Steel, MOU> + 100^<Wheat, MOU>
+ 200^<Wheat flour, MOU> + 400<Wages expenditure, MOU>
+ 95<Gross Profit, MOU> + 5<Reserve for depreciation, MOU>
(1c) fneeds[Industry] = fneeds[Products]
= 50^<Steel, MOU> + 100^<Wheat, MOU> + 200^<Wheat flour, MOU>
+ 300^<Bread, MOU> + 160^<Machinery, MOU>
(2a) fpro[Government] = 50<Government service, MOU>
(2b) fcost[Government]
= 50<Wages expenditure, MOU> + 1<Depreciation, MOU>
(2c) fneeds[Government] = 50^<Government service, MOU>

[Definition 6.5.9] Sector-Income Vector, Sector-Outcome Vector
[1] Sector-Income Vector: fincom[ω]
[2] Sector-Outcome Vector: foutcom[ω]
(1a) fincom[Industry]
= ¯{fsum[Gross Profit, Transfer of Gross Profit Set]
+ fsum[^Gross Profit, Transfer of Gross Profit Set]}

+ fsum[Interest earned, Industry Interest Earned Set]
+ fsum[Subsidy Income, Industry Subsidy Income Transaction Set]

(1b) foutcom[Industry]
= fsum[Interest expense, Industry Interest Transaction]
+ fsum[Taxes expense, Industry Tax Transaction]
+ fsum[Retained Earnings, Industry Retained Earnings Set]

For ω ∈ Industry,
(1c) fincom[ω]
= —{fsum[Gross Profit, {f[Transfer of Gross Profit: ω](ω)}]
+ fsum[^Gross Profit, {f[Transfer of Gross Profit: ω](ω)}]}
+ fsum[Interest earned, {f[Deposit Interest: ω](ω),
f[National Bond Interest: ω](ω)}]
+ fsum[Subsidy Income, {f[Government Subsidy: ω](ω)}]

(1d) foutcom[ω]
= fsum[Interest expense, {f[Loan Interest Expense: ω](ω)}]
+ fsum[Taxes expense, {f[Tax: ω](ω)}]
+ fsum[Retained Earnings, {f[Retained Earnings: ω](ω)}]

(2a) fincom[Bank]
= fsum[Interest earned, Bank Interest Earned Set]

(2b) foutcom[Bank]
= fsum[Interest expense, Bank Interest Expense Set]
+ fsum[Taxes expense, {f[Tax: Bank](Bank)}]
+ fsum[Retained Earnings, {f[Retained Earnings: Bank}](Bank}}]

(3a) fincom[Household]
= fsum[Wages earned, Household Wages Earned Transaction Set]
+ fsum[Interest earned, Household Interest Earned Set]

(3b) foutcom[Household]
= fsum[Consumption expenditure, Household Consumption Set]
+ fsum[Interest expense, Household Interest Expense Set]
+ fsum[Taxes expense, Household Taxes Expense Set]
+ fsum[Retained Earnings, Household Retained Earnings Set]

(4a) fincom[Government]
= fsum[Taxes revenue, Government Tax Revenue Set]
+ fsum[Central Bank Payment Income, Government Central Bank Payment Set]

(4b) foutcom[Government]
= fsum[Consumption expenditure, Government Consumption Expenditure Set] + fsum[Interest expense, Government Interest Expense Set]
+ fsum[Retained Earnings, Government Retained Earnings Set]

[Example 6.5.9] Example of Sector-Income Vector & Output Vector
fincom[Agriculture]
= 19<Gross Profit, MOU> + 0.9<Interest earned, MOU>
foutcom[Agriculture]
= 4<Interest expense, MOU> + 2<Taxes expense, MOU>
+ 13.9<Retained Earnings, MOU>
fincom[Flour-milling Industry]
= 19<Gross Profit, MOU> + 0.9<Interest earned, MOU>
foutcom[Flour-milling Industry]
= 2.4<Interest expense, MOU> + 2<Taxes expense, MOU>
+ 15.5<Retained Earnings, MOU>
fincom[Bread Industry]
= 19<Gross Profit, MOU> + 0.9<Interest earned, MOU>
foutcom[Bread Industry]
= 2.4<Interest expense, MOU> + 2<Taxes expense, MOU>
+ 15.5<Retained Earnings, MOU>
fincom[Steel Industry]
= 19<Gross Profit, MOU> + 0.9<Interest earned, MOU>
foutcom[Steel Industry]
= 2.4<Interest expense, MOU> + 2<Taxes expense, MOU>
+ 15.5<Retained Earnings, MOU>
fincom[Machinery Industry]
= 19<Gross Profit, MOU> + 0.9<Interest earned, MOU>
foutcom[Machinery Industry]
= 2.4<Interest expense, MOU> + 2<Taxes expense, MOU>
+ 15.5<Retained Earnings, MOU>
fincom[Industry]
= 95<Gross Profit, MOU> + 4.5<Interest earned, MOU>
foutcom[Industry]
= 13.6<Interest expense, MOU> + 10<Taxes expense, MOU>
+ 75.9<Retained Earnings, MOU>
fincom[Bank]
= 13.6<Interest earned, MOU>
foutcom[Bank]
= 13<Interest expense, MOU> + 3.6<Retained Earnings, MOU>
fincom[Central Bank]
= 3<Interest earned, MOU>
foutcom[Central Bank]
= 3<Central Bank Payment Expense, MOU>
fincom[Government]
= 50<Taxes revenue, MOU> + 1^<Gross Profit, MOU>
+ 3<Central Bank Payment Income, MOU>
foutcom[Government]
= 0.7<Interest expense, MOU> + 50<Consumption expenditure, MOU>
+ 1.3<Retained Earnings, MOU>

fincom[Household]
= 400<Wages earned, MOU> + 6.2<Interest earned, MOU>
foutcom[Household]
= 300<Consumption expenditure, MOU> + 40<Taxes expense, MOU>
+ 66.2<Retained Earnings, MOU>

[Definition 6.5.10] Sector-Accumulation Vector
For ω ∈ {Industry, Household, Government},
(1) Equipment investments Vector fEquipment investments[ω],
(2) Investments resource Vector fInvestments resource[ω],
(3) Deposits-investments difference[ω]
are defined as follows.

For ω ∈ {Industry, Bank, Household, Government, Central Bank}
(4) Current Assets increase Vector fCurrent Assets[ω]
(5) Current Liability increase Vector fCurrent Liability[ω],
(6) Cash Short and Over fCash Short and Over[ω]
are defined as follows.

(1)
fEquipment investments[Industry]
= fsum[^Machinery, Industry Equipment Investments Set]

fEquipment investments[Household]
= fsum[^Machinery, {f[Equipment investments Transfer: Machinery](Household)}]

fEquipment investments[Government]
= fsum[^Machinery, {f[Equipment investments Transfer: Machinery](Government)}]

(2)
fInvestments resource[Industry]
= fsum[Reserve for depreciation, Industry Depreciation Set]
+ fsum[Retained Earnings, Industry Retained Earnings Set]

fInvestments resource[Household]
= fsum[Reserve for depreciation, {f[Depreciation: Household](Household)}]
+ fsum[Retained Earnings, {f[Retained Earnings: Household](Household)}]

fInvestments resource[Government]
= fsum[Reserve for depreciation, {f[Depreciation: Government](Government)}]
+ fsum[Retained Earnings, {f[Retained Earnings: Government}](Government)}]

(3)
fDeposits-investments difference[ω]
= |fInvestments resource[ω]| − |fEquipment investments[ω]|

(4)

fCurrent Assets[Industry]
= —{∑{Flow[Cash](ω, t)<Cash, MOU>} + Flow[^Cash](ω, t)^<Cash, MOU>
+ Flow[Deposits](ω, t)<Deposits, MOU>
+ {Flow[^Deposits](ω, t)^<Deposits, MOU>
+ ∑ Flow[National Bonds](ω, t)<National Bonds, MOU>
+ Flow[^National Bonds](ω, t)^<National Bonds, MOU> | ω ∈ Industry}}

fCurrent Assets[Government]
= —{Flow[Cash](Government, t)<Cash, MOU>
+ Flow[^Cash](Government, t)^<Cash, MOU>}

fCurrent Assets[Household]
= —{Flow[Cash](Household, t)<Cash, MOU>
+ Flow[^Cash](Household, t)^<Cash, MOU>
+ Flow[Deposits](Household, t)<Deposits, MOU>
+ Flow[^Deposits](Household, t)^<Deposits, MOU>
+ Flow[National Bonds](Household, t)<National Bonds, MOU>
+ Flow[^National Bonds](Household, t)^<National Bonds, MOU>}

fCurrent Assets[Bank]
= —{Flow[Cash](Bank, t)<Cash, MOU>
+ Flow[^Cash](Bank, t)^<Cash, MOU>
+ ∑{Flow[Loans receivable: ω](Bank, t)<Loans receivable, MOU>
+ Flow[^Loans receivable: ω](Bank, t)^<Loans receivable, MOU>
| ω ∈ {Household} ∪ Industry}
+ Flow[Central Bank Reserve Fund Receivable](Bank, t)
<Reserve deposit receivable, MOU>
+ Flow[^Central Bank Reserve Fund Receivable](Bank, t)
^<Reserve deposit receivable, MOU>}

fCurrent Assets[Central Bank]
= —{Flow[Central BankLoans receivable](Central Bank, t)
<Loans receivable, MOU>
+ Flow[^Central BankLoans receivable](Central Bank, t)
^<Loans receivable, MOU>
+ Flow[National Bonds](Central Bank, t)<National Bonds, MOU>
+ Flow[^National Bonds](Central Bank, t)^<National Bonds, MOU>}

(5)

fCurrent Liability[Industry]
= —{∑{Flow[Loans payable](ω, t)<Loans payable, MOU> | ω ∈ Industry}
+ ∑{Flow[^Loans payable](ω, t)^<Loans payable, MOU> | ω ∈ Industry}}

fCurrent Liability[Household]
= —{Flow[Loans payable](Household, t)<Loans payable, MOU>
+ Flow[^Loans payable](Household, t)^<Loans payable, MOU>}

fCurrent Liability[Government]
= ¯{∑{Flow[National Bonds payable: ω](Government, t)
<National Bonds payable, MOU>
+ Flow[^National Bonds payable: ω](Government, t)
^<National Bonds payable, MOU> | ω ∈ {Household, Central Bank} ∪
Industry}}

fCurrent Liability[Bank]
= ¯{Flow[Central BankLoans payable](Bank, t)<Loans payable, MOU>
+ Flow[^Central BankLoans payable](Bank, t)^<Loans payable, MOU>
+ ∑{Flow[Deposits payable: ω](Bank, t)<Deposits payable, MOU>
+ Flow[^Deposits payable: ω](Bank, t)^<Deposits payable, MOU>
| ω ∈ {Household} ∪ Industry}}

fCurrent Liability[Central Bank]
= ¯{Flow[Central Bank Reserve Fund Payable](Central Bank, t)
<Reserve deposit payable, MOU>
+ Flow[^Central Bank Reserve Fund Payable](Central Bank, t)
^<Reserve deposit payable, MOU>
+ Flow[Central Bank note payable](Central Bank, t)
<Central Bank note payable, MOU>
+ Flow[^Central Bank note payable](Central Bank, t)
^<Central Bank note payable, MOU>}

(6)
fCash Short and Over[ω]
= |Project[Credit](fCurrent Assets[ω])
+ Project[Credit](fCurrent Liability[ω])|
− |Project[Debit](fCurrent Assets[ω])
+ Project[Debit](fCurrent Liability[ω])|

[Example 6.5.10] Example of a Sector-Accumulation Vector
(1) fEquipment investments[Industry]
= 140^<Machinery, MOU>

(2) fInvestments resource[Industry]
= 75.9<Retained Earnings, MOU> + 5<Reserve for depreciation, MOU>

(3) fDeposits-investments difference[Industry] = 140 − 80.9 = 59.1

(4) fCurrent Assets[Industry]
= ¯(332.4^<Cash, MOU> + 312.9<Cash, MOU>) + ¯(192.4^<Cash, MOU>
+ 172.9<Cash, MOU>) + ¯(72.4^<Cash, MOU> + 52.9<Cash, MOU>)
+ ¯(232.4^<Cash, MOU> + 212.9<Cash, MOU>) + ¯(124^<Cash, MOU>
+ 120.9<Cash, MOU>) + 18<Deposits, MOU> + 18<Deposits, MOU>
+ 18<Deposits, MOU> + 18<Deposits, MOU> + 18<Deposits, MOU>
= 81.1^<Cash, MOU> + 90<Deposits, MOU>

(5) fCurrent Liability[Industry]
= 12<Loans payable, MOU> + 12<Loans payable, MOU>
+ 12<Loans payable, MOU> + 12<Loans payable, MOU>
+ 12<Loans payable, MOU>
= 60<Loans payable, MOU>

(6) Cash Short and Over[Industry] = 90 – 81.1 – 60 = –51.1(Cash Short)

[Definition 6.5.11] Sector-Oversea Vector
(1) fImport[Industry] = fpro[Overseas Products]
(2) fExport[Industry] = fneeds[Product Export Transaction Set]

[Example 6.5.11] Example of Sector-Oversea Vector
(1) fImport[Industry] = 10<Steel, MOU>
(2) fExport[Industry] = 10<Machinery, MOU>

6.5.2 Sector-National Economy Field: Stock Quantities

[Definition 6.5.12] Sector-Initial, Final Balance Sheet
f[Final Stock](ω, t)
= —{f[st, Initial](ω, t) + f[Interim flow](ω, t) | $\omega \in \Omega_1$ – {Oversea}}
f[Final Stock](Industry, t)
= —{f[st, Initial](Industry, t) + f[Interim flow](Industry, t)}

[Example 6.5.12] Example of Sector-Interim flow, Final Balance Sheet
(1) Interim Flow
f[Interim flow](Agriculture, t)
= 124^<Cash, MOU> + 120.9<Cash, MOU> + 20<Machinery, MOU>
+ 13.9<Retained Earnings, MOU> + 18<Deposits, MOU>
+ 20<Loans payable, MOU> + 1<Reserve for depreciation, MOU>

f[Interim flow](Flour-milling Industry, t)
= 232.4^<Cash, MOU> + 212.9<Cash, MOU> + 30<Machinery, MOU>
+ 15.5<Retained Earnings, MOU> + 18<Deposits, MOU>
+ 12<Loans payable, MOU> + 1<Reserve for depreciation, MOU>

f[Interim flow](Bread Industry, t)
= 332.4^<Cash, MOU> + 312.9<Cash, MOU> + 30<Machinery, MOU>
+ 15.5<Retained Earnings, MOU> + 18<Deposits, MOU>
+ 12<Loans payable, MOU> + 1<Reserve for depreciation, MOU>

f[Interim flow](Steel Industry, t)
= 72.4^<Cash, MOU> + 52.9<Cash, MOU> + 30<Machinery, MOU>
+ 15.5<Retained Earnings, MOU> + 18<Deposits, MOU>
+ 12<Loans payable, MOU> + 1<Reserve for depreciation, MOU>

f[Interim flow](Machinery Industry, t)
= 192.4^<Cash, MOU> + 172.9<Cash, MOU> + 30<Machinery, MOU>

+ 15.5<Retained Earnings, MOU> + 18<Deposits, MOU>
+ 12<Loans payable, MOU> + 1<Reserve for depreciation, MOU>

f[Interim flow](Household, t)
= 460^<Cash, MOU> + 406.2<Cash, MOU>
− 66.2<Retained Earnings, MOU> + 110<Deposits, MOU>
+ 10<National Bonds, MOU>

f[Interim flow](Government, t)
= 60.7^<Cash, MOU> + 63<Cash, MOU>
+ 10<Machinery, MOU> + 1.3<Retained Earnings, MOU>
+ 1<Reserve for depreciation, MOU> + 10<National Bonds payable, MOU>

f[Interim flow](Bank, t)
= 313.6<Cash, MOU> + 131^<Cash, MOU> + 0.6<Retained Earnings, MOU>
+ 200<Deposits payable, MOU> + 68<Loans receivable, MOU>
+ 50<Reserve deposit receivable, MOU> + 100<Loans payable, MOU>

f[Interim flow](Central Bank, t)
= 53<Cash, MOU> + 103^<Cash, MOU>
+ 50<Reserve deposit payable, MOU> + 100<Loans receivable, MOU>
+ 10<National Bonds, MOU> + 10^<National Bonds, MOU>

f[Interim flow](Industry, t)
= 872.5<Cash, MOU> + 953.6^<Cash, MOU> + 140<Machinery, MOU>
+ 75.9<Retained Earnings, MOU> + 90<Deposits, MOU>
+ 68<Loans payable, MOU> + 5<Reserve for depreciation, MOU>

—f[Interim flow](Industry, t)
= 81.1^<Cash, MOU> + 140<Machinery, MOU>
+ 75.9<Retained Earnings, MOU> + 90<Deposits, MOU>
+ 68<Loans payable, MOU> + 5<Reserve for depreciation, MOU>

Flow is always balanced.
For example in the case of agriculture,
|120.9<Cash, MOU> + 20<Machinery, MOU> + 18<Deposits, MOU>|
= |124^<Cash, MOU> + 13.9<Retained Earnings, MOU>
+ 20<Loans payable, MOU> + 1<Reserve for depreciation, MOU>| = 158.9

(2) Final Stock
For the calculation we use the initial stock given in the Example 6.4.1.

f[Final Stock](ω, t)
= —{f[Initial Stock](ω, t) + f[Interim flow](ω, t)}

f[Final Stock](Agriculture, t)
= 146.9<Cash, MOU> + 70<Machinery, MOU> + 100<Capital stock, MOU>
+ 113.9<Retained Earnings, MOU> + 18<Deposits, MOU>
+ 20<Loans payable, MOU> + 1<Reserve for depreciation, MOU>

f[Final Stock](Flour-milling Industry, t)
= 130.5<Cash, MOU> + 280<Machinery, MOU>
+ 300<Capital stock, MOU> + 115.5<Retained Earnings, MOU>
+ 18<Deposits, MOU> + 12<Loans payable, MOU>
+ 1<Reserve for depreciation, MOU>

f[Final Stock](Bread Industry, t)
= 330.5<Cash, MOU> + 180<Machinery, MOU>
+ 200<Capital stock, MOU> + 315.5<Retained Earnings, MOU>
+ 18<Deposits, MOU> + 12<Loans payable, MOU>
+ 1<Reserve for depreciation, MOU>

f[Final Stock](Steel Industry, t)
= 80.5<Cash, MOU> + 130<Machinery, MOU> + 100<Capital stock, MOU>
+ 115.5<Retained Earnings, MOU> + 18<Deposits, MOU>
+ 12<Loans payable, MOU> + 1<Reserve for depreciation, MOU>

f[Final Stock](Machinery Industry, t)
= 80.5<Cash, MOU> + 280<Machinery, MOU> + 250<Capital stock, MOU>
+ 165.5<Retained Earnings, MOU> + 18<Deposits, MOU>
+ 12<Loans payable, MOU> + 1<Reserve for depreciation, MOU>

f[Final Stock](Household, t)
= 446.2<Cash, MOU> + 500<Machinery, MOU> + 400<Capital stock, MOU>
+ 666.2<Retained Earnings, MOU> + 110<Deposits, MOU>
+ 10<National Bonds, MOU>

f[Final Stock](Government, t)
= 502.3<Cash, MOU> + 160<Machinery, MOU> + 200<Capital stock, MOU>
+ 302.3<Retained Earnings, MOU> + 1<Reserve for depreciation, MOU>
+ 160<National Bonds payable, MOU>

f[Final Stock](Bank, t)
= 682.6<Cash, MOU> + 60<Machinery, MOU> + 200<Capital stock, MOU>
+ 300.6<Retained Earnings, MOU> + 200<Deposits payable, MOU>
+ 68<Loans receivable, MOU> + 350<Reserve deposit receivable, MOU>
+ 450<Loans payable, MOU>

f[Final Stock](Central Bank, t)
= 0<Cash, MOU> + 1000<Capital stock, MOU>
+ 350<Reserve deposit payable, MOU>
+ 2450<Central Bank note payable, MOU> + 450<Loans receivable, MOU>
+ 3200<Gold, MOU> + 150<National Bonds, MOU>

6.6 Integrated National Economic Field

We will give integrated field quantities for a national economy.

6.6.1 Integrated National Economy Field: Flow Quantities

[Definition 6.6.1]
Domestic Integrated cost Vector: fDomestic cost
Domestic Integrated needs Vector: fDomestic needs
(1) fDomestic cost = fcost[Industry] + fcost[Government]
(2) fDomestic needs = fneeds[Industry] + fneeds[Government]

[Example 6.6.1] Examples of fDomestic Cost and fDomestic Needs
(1) fDomestic cost = fcost[Industry] + fcost[Government]
= 50^<Steel, STU> + 100^<Wheat, MOU> + 200^<Wheat flour, MOU>
+ 450<Wages expenditure, MOU> + 95<Gross Profit, MOU>
+ 6<Reserve for depreciation, MOU>
(2) fDomestic needs = fneeds[Industry] + fneeds[Government]
= 50^<Steel, STU> + 100^<Wheat, MOU> + 200^<Wheat flour, MOU>
+ 300^Bread + 160^<Machinery, MOU> + 50^<Government service, MOU>

[Definition 6.6.2]
Domestic Integrated production Vector: fDomestic pro,
Integrated output Vector: fDomestic con
(1) fDomestic pro
= fsum[Wages expenditure, Industry Wage Payment Transaction Set]
+ fsum[Gross Profit, Transfer of Gross Profit Set]
+ fsum[Reserve for depreciation, Industry Depreciation Set]
+ fsum[Wages expenditure, Government Wage Payment Transaction Set]
+ fsum[Reserve for depreciation, Government Depreciation Set]
= ‾(fDomestic cost + ∑{^$f_U(\omega)$ | $\omega \in \{a_1, \ldots, a_n\}$}
+ ∑{^fsum[^e, Government Production Transaction Set] | e ∈ Products})

(2) fDomestic con
= ‾(fDomestic needs + ^fneeds[Production Transaction Set])

[Definition 6.6.3]
National Income Vector: fNational Income
National Outcome Vector: fNational Outcome
(1) fNational income
= ‾{fsum[Gross Profit, Transfer of Gross Profit Set]
+ fsum[^Gross Profit, Transfer of Gross Profit Set]}
+ fsum[Wages earned, Wage Payment Transaction Set]

(2) fNational outcome
= fsum[Retained Earnings, Industry Retained Earnings Set]
+ fsum[Retained Earnings, Bank Retained Earnings Set]
+ fsum[Consumption expenditure, Household Consumption Set]
+ fsum[Retained Earnings, Household Retained Earnings Set]
+ fsum[Consumption expenditure, Government Consumption Expenditure Set]
+ fsum[Retained Earnings, Government Retained Earnings Set]

[Definition 6.6.4] National Accumulation Vector
National Stock investments Vector: fNational Stock investments
National Investments resource Vector: fNational Investments resource

(1) fNational Stock investments
= ∪ {fStock investments[ω] | ω ∈ Ω_1 − {Overseas}}
(2) fNational Investments resource
= ∪ {fInvestments resource[ω] | ω ∈ Ω_1 − {Overseas}}

[Definition 6.6.5] Overseas Operating Transaction Vector

fImport = fpro[Overseas Product]
fExport = fneeds[Product Export Transaction Set]
National operating surplus = |fExport| − |fImport|
Note: In this model we do not treat capital flow, only consider transactions of products from and to abroad.

[Example 6.6.2] Example of an Overseas Operating Transaction Vector

(1) fImport = fpro[Overseas Product] = 10<Steel, MOU>
(2) fExport = fneeds[Product Export Transaction Set] = 10^<Machinery, MOU>
National operating surplus = |fExport| — |fImport| = 0

6.6.2 Integrated National Economic Field: Stock Quantities

[Definition 6.6.6]
National Initial Balance Sheet Vector,
National Final Balance Sheet Vector

(1) f[st, Initial](Nation, t)
= ¯(Σ{f[st, Initial](ω, t) | ω ∈ Ω_1 − {Oversea}})

(2) f[Interim flow](Nation, t)
= ¯(Σ{f[Interim flow](ω, t) | ω ∈ Ω_1 − {Oversea}})

(3) f[Final Stock](Nation, t)
= ¯(f[st, Initial](Nation, t) + f[Interim flow](Nation, t))
= ¯(Σ{f[Final Stock](ω, t) | ω ∈ Ω_1 − {Oversea}})

[Example 6.6.3]

(1) f[st, Initial](Nation, t)
= ¯(Σ{f[st, Initial](ω, t) | ω ∈ Ω_1 − {Oversea}})
= 2400<Cash, MOU> + 2750<Capital stock, MOU>
+ 1500<Machinery, MOU> + 1950<Retained Earnings, MOU>
+ 150<National Bonds, MOU> + 150<National Bonds payable, MOU>
+ 300<Reserve deposit receivable, MOU>
+ 300<Reserve deposit payable, MOU> + 350<Loans receivable, MOU>
+ 350<Loans payable, MOU> + 2400<Central Bank note payable, MOU>
+ 3200<Gold, MOU>

(2) f[Interim flow](Nation, t)
= ¯(∑{f[Interim flow](ω, t) |ω ∈ Ω_1 – {Oversea}})
= 0<Cash, MOU> + 144<Retained Earnings, MOU>
+ 200<Deposits, MOU> + 10<National Bonds, MOU>
+ 200<Deposits payable, MOU> + 10<National Bonds payable, MOU>
+ 68<Loans receivable, MOU> + 68<Loans payable, MOU>
+ 150<Machinery, MOU> + 6<Reserve for depreciation, MOU>

(3) f[Final Stock](Nation, t)
= ¯(f[st, Initial](Nation, t) + f[Interim flow](Nation, t))
= ¯(∑{f[Final Stock](ω) | ω ∈ Ω_1 – {Oversea}})
= 2400<Cash, MOU> + 1750<Capital stock, MOU>
+ 1650<Machinery, MOU> + 2094<Retained Earnings, MOU>
+ 200<Deposits, MOU> + 160<National Bonds, MOU>
+ 200<Deposits payable, MOU> + 160<National Bonds payable, MOU>
+ 68<Loans receivable, MOU> + 68<Loans payable, MOU>
+ 6<Reserve for depreciation, MOU>
+ 2400<Central Bank note payable, MOU> + 3200<Gold, MOU>

6.6.3 Basic Identity Equation among Macro Economic Variables

Now we prove the following identity equation from our formulation.
(1) Y[National Economy] = S[National Economy] + C[National Economy]
(2) Y[National Economy] = I[National Economy] + C[National Economy]
+ EX[National Economy] – IM[National Economy]
The identity equations are famous but are treated as a definition. In the theory of systems of national accounts it is recognized as a fully-articulated system. We prove this from a bottom up construction point of view.

[Cor. 6.6.1]
(1)
For retained earnings the following formulas hold.
(1-A) Wages[Household](t)
= ∑{Wages[ω](t) | ω ∈ Industry ∪ {Government}}
(1-B) Subsidy[Government](t)
= ∑{Subsidy[ω](t) | ω ∈ Industry ∪ {Household, Bank}}
(1-C) NBondInt[Government](t)
= ∑{NBondInt[ω](t) | ω ∈ Industry ∪ {Household}}
(1-D) LoanIntIn[Bank](t) = ∑{LoanInt[ω](t) | ω ∈ Industry ∪ {Household}}
(1-E) SaveIntOut[Bank](t) = ∑{SaveInt[ω](t) | ω ∈ Industry ∪ {Household}}
(1-F) LoanIntOut[Bank](t) = CBLoanInt(t)
(1-G) GovReceipt(t) = CBLoanInt(t)
(1-H) TaxRevenue(t)
= ∑{CompTax[ω](t) | ω ∈ {Bank} ∪ Industry} + IncomeTax(t)

(2)
For consumption the following formulas hold.
(2-A) SellPrice[Bread](t) = HouseDomCons(t)
(2-B) C(Household) = Import[Bread](Household, t) + HouseDomCons(t)
(2-C) Let agent ω use domestic materials then,
SellPrice[em(ω)](t) = SellPrice[ep(α)](t),
SellPrice[em(Flour-milling Industry)](t) = SellPrice[ep(Agriculture)](t),
SellPrice[em(Bread Industry)](t) = SellPrice[ep(Flour-milling Industry)](t),
SellPrice[em(Machinery Industry)](t) = SellPrice[ep(Steel Industry)](t).

(3)
The following equations for ω ∈ Industry hold from the definition.
I[Product Stock Investment: ep(ω)](ω)
= ProPrice[ep(ω)](t) − SellPrice[ω](t) − Export[ep(ω)](ω, t)
Where ProPrice[ep(ω)](t) = Use[em(ω)](t) + ValueAdd[ω](t).
Use[em(ω)](t) = 0 for ω ∈ {Agriculture, Steel Industry}.
I[Material Stock Investment: em(ω)](ω)
= SellPrice[em(ω)](t) + Import[em(ω)](ω, t) − Use[em(ω)](t)
for ω ∈ {Flour-milling Industry, Bread Industry, Machinery Industry}.

(4)
Y(Government) = C(Government) = Wages[Government](t)

Proof:
It is clear from the definition. Q.E.D.

[Theorem 6.6.1] Basic Identity Equations of National Economy
(1) Y[National Economy] = S[National Economy] + C[National Economy]
(2) Y[National Economy] = I[National Economy] + C[National Economy]
+ EX[National Economy] − IM[National Economy]

Proof:
S[National Economy]
= Σ{S(ω) | ω ∈ Industry ∪ {Household, Government, Bank}}
= Σ{S[Retained Earnings](ω) | ω ∈ Industry ∪ {Household, Government, Bank}}
+ Σ{S[Depreciation](ω) | ω ∈ Industry ∪ {Household, Government}}
= Σ{|Project[<Retained Earnings, MOU>](f[Retained Earnings: ω](ω))| | ω ∈ Industry ∪ {Household, Government, Bank}}
− Σ{|Project[^<Retained Earnings, MOU>](f[Retained Earnings: ω](ω))| | ω ∈ Industry ∪ {Household, Government, Bank}}
+ Σ{|Project[<Reserve for depreciation, MOU>](f[Depreciation: ω](ω))| | ω ∈ Industry ∪ {Household, Government}}

C[National Economy]
= C(Household) + C(Government)
= |Project[<Consumption expenditure, MOU>]
(f[Household_Consumption expenditure](Household))|

+ |Project[<Consumption expenditure, MOU>]
(f[Government Service Expense](Government))|
= HouseDomCons(t) + Import[Bread](Household, t) + Wages[Government](t)

Y[National Economy]
= ∑{Y(ω) | ω ∈ Industry ∪ {Government}}
= ∑{|Project[<Value added, MOU>](f[Production: ep(ω)](ω)
+ f[Value added_Product Stock Adjustment](ω))|
− |Project[^<Value added, MOU>]
(f[Value added_Product Stock Adjustment](ω))| | ω ∈ Industry}
+ |Project[<Value added, MOU>](f[Government Service Production]
(Government))|
= ∑{ValueAdd[ω](t) + ValAdjustUp[ω](t)
− ValAdjustDown[ω](t) | ω ∈ Industry} + Wages[Government](t)

Let ω ∈ {Flour-milling Industry, Bread Industry, Machinery Industry}.
I(ω) = I[Equipment investments](ω) + I[Product Stock Investment](ω)
+ I[Material Stock Investment](ω)
+ I[Value added_Product Stock Adjustment](ω)
Let ω ∈ {Steel Industry, Agriculture}.
I(ω) = I[Equipment investments](ω) + I[Product Stock Investment](ω)
+ I[Value added_Product Stock Adjustment](ω)
Let ω ∈ {Government, Household}.
I(ω) = I[Equipment investments](ω)

Then

I[National Economy]
= ∑{I[Material Stock Investment](ω) | ω ∈ {Flour-milling Industry, Bread Industry, Machinery Industry}}
+ ∑{I[Product Stock Investment](ω) | ω ∈ Industry}
+ ∑{I[Value added_Product Stock Adjustment](ω) | ω ∈ Industry}
+ ∑{I[Equipment investments](ω) | ω ∈ Industry ∪ {Household, Government}}

EX[National Economy]
= ∑{EX[ep(ω)](ω) | ω ∈ Industry}
= ∑{Export[ep(ω)](ω, t) | ω ∈ Industry}
IM[National Economy]
= ∑{IM[em(ω)](ω) | ω ∈ {Flour-milling Industry, Bread Industry, Machinery Industry}}
+ ∑{IM[Machinery](ω) | ω ∈ Industry ∪ {Government, Household}}
+ IM[Bread](Household)
= ∑{Import[em(ω)](ω, t) | ω ∈ {Flour-milling Industry, Bread Industry, Machinery Industry}}
+ ∑{Import[Machinery](ω, t) | ω ∈ Industry ∪ {Government, Household}}
+ Import[Bread](Household, t)

(1) We show Y[National Economy] = C[National Economy] + S[National Economy] at first.

Y[National Economy]
= ∑{ValueAdd[ω](t) + ValAdjustUp[ω](t)
− ValAdjustDown[ω](t) | ω ∈ Industry} + Wages[Government](t)

C[National Economy] + S[National Economy]
= HouseDomCons(t) + Import[Bread](Household, t) + Wages[Government](t)
+ ∑{SaveInt[ω](t) | ω ∈ Industry} + ∑{NBondInt[ω](t) | ω ∈ Industry}
+ ∑{Subsidy[ω](t) | ω ∈ Industry} + ∑{ValueAdd[ω] | ω ∈ Industry}
+ ∑{ValAdjustUp[ω](t) | ω ∈ Industry} + Wages[Household](t)
+ SaveInt[Household](t) + NBondInt[Household](t)
+ Subsidy[Household](t) + TaxRevenue(t) + GovReceipt(t)
+ LoanIntIn[Bank](t) + Subsidy[Bank](t)
+ ∑{Dep[ω](t) | ω ∈ Industry ∪ {Government, Household}}
− {∑{LoanInt[ω](t) | ω ∈ Industry} + ∑{CompTax[ω](t) | ω ∈ Industry}
+ ∑{Wages[ω](t) | ω ∈ Industry} + ∑{Dep[ω](t) | ω ∈ Industry}
+ ∑{ValAdjustDown[ω](t) | ω ∈ Industry} + LoanInt[Household](t)
+ IncomeTax(t) + HouseDomCons(t) + Import[Bread](Household, t)
+ Dep[Household](t) + Dep[Government](t) + Wages[Government](t)
+ Subsidy[Government](t) + NBondInt[Government](t)
+ SaveIntOut[Bank](t) + LoanIntOut[Bank](t) + IncomeTax(t)}

The offset calculation is given by the definition of Wages[Household](t), Subsidy[Government](t), NBondInt[Government](t), LoanIntIn[Bank](t), SaveIntOut[Bank](t), TaxRevenue(t) and CBLoanInt(t).

C[National Economy] + S[National Economy]
= HouseDomCons(t)
+ Import[Bread](Household, t) + Wages[Government](t)
+ SaveIntOut[Bank](t) + NBondInt[Government](t)
+ Subsidy[Government](t) + ∑{ValueAdd[ω] | ω ∈ Industry}
+ ∑{ValAdjustUp[ω](t) | ω ∈ Industry} + Wages[Household](t)
+ TaxRevenue(t) + GovReceipt(t) + LoanIntIn[Bank](t)
+ ∑{Dep[ω](t) | ω ∈ Industry ∪ {Government, Household}}
− {LoanIntIn[Bank](t) + TaxRevenue(t) + Wages[Household](t)
+ ∑{Dep[ω](t) | ω ∈ Industry} + ∑{ValAdjustDown[ω](t) | ω ∈ Industry}
+ HouseDomCons(t) + Import[Bread](Household, t)
+ Dep[Household](t) + Dep[Government](t) + Subsidy[Government](t)
+ NBondInt[Government](t) + SaveIntOut[Bank](t) + CBLoanInt(t)}
= Wages[Government](t) + ∑{ValueAdd[ω] | ω ∈ Industry}
+ ∑{ValAdjustUp[ω](t) | ω ∈ Industry}
− {∑{ValAdjustDown[ω](t) | ω ∈ Industry}}
= Y[National Economy]

(2) Y[National Economy] = I[National Economy] + C[National Economy]
I[National Economy]
= ∑{I[Product Stock Investment](ω) | ω ∈ Industry}
+ ∑{I[Material Stock Investment](ω) | ω ∈ {Flour-milling Industry, Bread Industry, Machinery Industry}}
+ ∑{I[Value added_Product Stock Adjustment](ω) | ω ∈ Industry}
+ ∑{I[Equipment investments](ω) | ω ∈ Industry ∪ {Household, Government}}
= ∑{ValueAdd[ω](t) | ω ∈ {Agriculture, Steel Industry}}
− ∑{SellPrice[ep(ω)](t) | ω ∈ {Agriculture, Steel Industry}}
− ∑{Export[ep(ω)](ω, t) | ω ∈ {Agriculture, Steel Industry}}
+ ∑{Use[em(ω)](t)](ω, t) | ω ∈ {Flour-milling Industry, Bread Industry}}
+ ∑{ValueAdd[ω](t) | ω ∈ {Flour-milling Industry, Bread Industry}}
− ∑{SellPrice[ep(ω)](t) | ω ∈ {Flour-milling Industry, Bread Industry}}
− ∑{Export[ep(ω)](ω, t) | ω ∈ {Flour-milling Industry, Bread Industry}}
+ Use[Steel](t) + ValueAdd[Machinery Industry](t)
− ∑{InvPrice[ω](t) | ω ∈ Industry ∪ {Household, Government}}
− Export[Machinery](Machinery Industry, t)
+ ∑{SellPrice[em(ω)](t) | ω ∈ {Flour-milling Industry, Bread Industry, Machinery Industry}}
+ ∑{Import[em(ω)](ω, t) | ω ∈ {Flour-milling Industry, Bread Industry, Machinery Industry}}
− ∑{Use[em(ω)](t) | ω ∈ {Flour-milling Industry, Bread Industry, Machinery Industry}}
+ ∑{ValadjustUp[ω](t) | ω ∈ Industry}
− ∑{ValAdjustDown[ω](t) | ω ∈ Industry}
+ ∑{InvPrice[ω](t) | ω ∈ Industry ∪ {Household, Government}}
+ ∑{Import[Machinery](ω, t) | ω ∈ Industry ∪ {Household, Government}}

Offset calculation is given by the definition of SellPrice[em(Flour-milling Industry)](t) = SellPrice[ep(Agriculture)](t), SellPrice[Bread](t) = HouseDomCons(t).

C[National Economy] + I[National Economy]
= Wages[Government](t)
+ ∑{ValueAdd[ω](t) | ω ∈ {Agriculture, Steel Industry}}
+ ∑{ValueAdd[ω](t) | ω ∈ {Flour-milling Industry, Bread Industry}}
+ ValueAdd[Machinery Industry](t)
+ ∑{ValadjustUp[ω](t) | ω ∈ Industry}
− ∑{ValAdjustDown[ω](t) | ω ∈ Industry}
+ ∑{Use[em(ω)](t)](ω, t) | ω ∈ {Flour-milling Industry, Bread Industry}}
+ Use[Steel](t) − ∑{Use[em(ω)](t) | ω ∈ {Flour-milling Industry, Bread Industry, Machinery Industry}}
+ ∑{SellPrice[em(ω)](t) | ω ∈ {Flour-milling Industry, Bread Industry, Machinery Industry}}
+ HouseDomCons(t)

– ∑{SellPrice[ep(ω)](t) | ω ∈ {Agriculture, Steel Industry}}
– ∑{SellPrice[ep(ω)](t) | ω ∈ {Flour-milling Industry, Bread Industry}}
+ ∑{Import[em(ω)](ω, t) | ω ∈ {Flour-milling Industry, Bread Industry, Machinery Industry}}
+ ∑{Import[Machinery](ω, t) | ω ∈ Industry ∪ {Household, Government}}
+ Import[Bread](Household, t)
+ ∑{InvPrice[ω](t) | ω ∈ Industry ∪ {Household, Government}}
– ∑{InvPrice[ω](t) | ω ∈ Industry ∪ {Household, Government}}
– ∑{Export[ep(ω)](ω, t) | ω ∈ {Flour-milling Industry, Bread Industry}}
– ∑{Export[ep(ω)](ω, t) | ω ∈ {Agriculture, Steel Industry}}
– Export[Machinery](Machinery Industry, t)
= Wages[Government](t) + ∑{ValueAdd[ω](t) + ValAdjustUp[ω](t)
– ValAdjustDown[ω](t) | ω ∈ Industry}
+ IM[National Economy] – EX[National Economy]
= Y[National Economy] + IM[National Economy] – EX[National Economy]

Q.E.D.

6.7 Conclusion

In this chapter we have formulated an algebraic reconstruction of macroscopic variables from the microscopic variables of economic transactions. Macroscopic variables include not only simple macro economic variables such as Y, I, C and S but also complex statistical variables used in the SNA (System of National Accounts). The Systems of National Accounts 1968 were revised as the SNA in 1993. The SNA 1993 provides a new standard set of coherent, consistent and integrated macroeconomic accounts, balance sheets and tables depending not on algebraic concepts but on integrated statistical concepts.

We have only reformulated the macroeconomic variables of a virtual economy with our algebraic framework. We need to extend our model to include the total statistics of the SNA. It is not a difficult work in our framework.

Then we can reconstruct not only SNA but also a macro econometric model as a nonlinear agent-based model from algebraic and bottom-up points of view.

Chapter 7
Gaming Simulation and the Dynamics of a Virtual Economy

Aim of This Chapter

In this chapter we analyze economic systems as agent based bottom up models. For this purpose we use the frameworks introduced in the previous chapters, such as a small national economy called a virtual economy, exchange algebra for state space descriptions and gaming simulation. We construct and analyze a dynamic agent-based and gaming simulation model.

For this purpose we will introduce a gaming simulation of a virtual economy. We analyze the model with a hybrid approach using a machine agent model and a human gaming simulation model. A machine agent model will be used to investigate an effective growth path by genetic algorithm (GA). We can find an effective growth path by using GA with complex boundary conditions.

7.1 Introduction

In this chapter we construct a Simulation and Gaming model of a Virtual Economy. A Virtual Economy consists of nine agents such as Agriculture, a Milling Industry, a Bread Industry (Bakery), Steel Manufacture, a Machinery Industry, Government, Households, a Bank and Central Bank. Algebraic abstraction of a bookkeeping system called exchange algebra is introduced to describe micro economic exchange. The economic state of each agent is also described by this algebra. Exchange algebra is an extension of accounting vector space as we have shown in Chapter 5 [Deguchi, 1986, 1990]. By using this algebra we describe the systemic properties of economic exchange and the properties of an economic field, which gives a formal model for the SNA (System of National Accounts) as is shown in Chapter 6.

In this virtual economy Agriculture grows wheat. The milling industry grinds wheat flour from the wheat. The bread industry (bakery) bakes bread from flour. Steel manufacture forges steel. The machinery industry makes machinery from steel. Households purchase and consume bread and also supply workers and receive a wage.

A machine is purchased by each industry as an investment and is also purchased by the Government or households. Machinery depreciates in value. Population increases by a scenario. The Government can issue National Bonds. The Central Bank issues a bank note and decides the official bank rate. Households and industry deposit money in a bank. A bank lends money.

7.2 Assumption of Virtual Economy Gaming

In the virtual economy gaming players act out roles as government, agriculture and so on. This virtual economy becomes a multi agent model of an economic system of a country. In the economy players or machine agents act as decision makers. The game needs some basic assumptions. For example we have five products and one currency in this economy with proper units for the products and currency as follows: "MOU" stands for a money unit such as the dollar. "WHU" stands for a wheat unit. "FLU" stands for a flour unit. "BRU" stands for a bread unit. "STU" stands for a steel unit and "MAU" stands for a machine unit. A machine is regarded as a house if a household purchases it.

We have constructed an agent-based simulation model for this economy with the following additional scenario and assumptions.

(1) Initial Price System for Five Products

Wheat	0.2 MOU/WHU	Flour	0.5 MOU/FLU
Bread	1 MOU/BRU	Steel	6.25 MOU/STU
Machine	10 MOU/MAU		

(2) Initial Parameter of Tax and Interest

Income Tax Rate	20%	Company Tax Rate	10%
Deposits Interest	1%	Loan Interest	3%

(3) Initial Employment and Wages

Initial Employment		Total Wages	
Agriculture	60 workers	Wages Expense	80 MOU
Milling Industry	60 workers	Wages Expense	80 MOU
Bread Industry	60 workers	Wages Expense	80 MOU
Steel Manufacture	20 workers	Wages Expense	20 MOU
Machine Industry	65 workers	Wages Expense	90 MOU
Government	35 workers	Wages Expense	50 MOU

(4) Increasing Population

Population of this small economy increases from 300 by 10% in each term as follows.

$$x(t+1) = \mathrm{int}(x(t) \times 1.1)$$

1 term	300.00	2 term	330.00	3 term	363.00
4 term	399.00	5 term	438.00	6 term	481.00
7 term	529.00	8 term	581.00	9 term	639.00
10 term	702.00				

(5) Basic Food Consumption
Minimum food consumption is assumed to be 1 BRU per person in a term. Persons consume food and invest in houses. The ratio depends on a player's preference.

(6) Production Function
We introduce three types of production functions as follows.

(6-1) Maximum Production per one Machine

Agricultural Machine:	110<Wheat, WHU>/an Agricultural Machine
Milling Machine:	20<Flour, FLU>/a Milling Machine
Bread Baking Machine:	25<Bread, BRU>/a Bread Baking Machine
Steel Making Machine:	1<Steel, STU>/a Steel Making Machine
Machine Tool:	1<Machinery, MAU>/a Machine Tool

(6-2) The Number of Workers to Use a Machine

Agricultural Machine:	10 workers/an Agricultural Machine
Milling Machine:	2 workers/a Milling Machine
Bread Baking Machine:	3 workers/a Bread Baking Machine
Steel Making Machine:	2 workers/a Steel Making Machine
Machine Tool:	2 workers/a Machine Tool

(6-3) Input Output Ratio from Materials to Product

Agricultural Machine:	No Materials
Milling Machine:	0.8 FLU/WHU
Bread Baking Machine:	(3/4) = 0.75 BRU/FLU
Steel Making Machine:	No Materials
Machine Tool:	2 MAU/STU

In our research program we will develop this agent-based modeling of a virtual economy and introduce three types of dynamic models for virtual economy gaming and develop a mixed model of software agents and human agents. However we will introduce gaming simulation models only with human players at this stage. We want to clarify the variety of the dynamics of economic systems, in which we represent not only the boom and bust cycles of the economy but also the different varieties of economic policy failure and crisis. Gaming simulation gives rise to a rich diversity of economic conditions and allows us many useful insights into the behavior of an economy.

7.3 Three Types of Gaming Simulation for a Virtual Economy

7.3.1 Dictator's View Model

We introduce three types of dynamic models for this virtual economy gaming. The first is called a dictator's view model. In this model a single player acting as a dictator determines all decisions for transactions among agents of the virtual economy in a term. In the model decisions are taken progressively at each term. Table 7.1 shows an example of decision-making items for a player in a term.

7.3.2 Bird's Eye View Model

The second is called a bird's eye view model or the God's Eye Model across the terms. A central authority player has a bird's eye view across several terms and can observe a total period of economic development in this economy and decide throughout the terms ways to achieve the chosen purpose of this economy.

The purpose is shown on a macro social index such as (1) average food consumption per person, (2) GDP per person and (3) the average number of residents in a house. Food consumption is used as a consumption utility in this economy.

In this model some decisions are made automatically depending on several hidden almost rational decision-making rules. A player decides across time in this model with a bird's eye view and can observe a total period of economic development and decide across the terms his strategy for achieving his aim in the economy.

Table 7.2 shows an example of the institutional parameters for the subsidy and national bond policy. The decision is shown as true or false. At first this part is also investigated under the centralized condition, but later with a decentralized and multi agent mode analysis.

Table 7.3 shows an example of the capital investment of each agent in each term. In the bird's eye view model a player determines all capital investment across the terms within the internal restrictions of this economy. We investigated this part by using genetic algorithm in section 7.7. The second column shows the machines available in each term. The machines are sold to each industrial sector as an investment. The numbers of available machines in each term change dynamically in response to the history of past investment policy. Table 7.4 shows other management and political parameters. In this example we fixed these parameters.

A player can observe economic development across the terms while he changes these parameters. A player can set up different goals for optimizing depending on focusing social indexes such as the numbers of residents per house, GDP per person and consumption per person.

TABLE 7.1 Decision Making for Dictator's View Model

	Decision Making for Dictater's View Model: 2nd Term								
	Agri.	Flou.	Bake.	Steel	Mach.	House	Gov.	CB	Bank
Income Tax Rate	*	*	*	*	*	*	0.1	*	*
Corporate Tax Rate	*	*	*	*	*	*	0.2	*	*
Natinoal Bond Rate	*	*	*	*	*	*	0.01	*	*
Official Bank Rate	*	*	*	*	*	*	*	0.01	*
Deposits in CB	*	*	*	*	*	*	*	0	0
Withdraws from CB	*	*	*	*	*	*	*	0	0
Loan from CB	*	*	*	*	*	*	*	1000	1000
Refund to CB	*	*	*	*	*	*	*	0	0
Subsidy Allocations	0	0	10	10	10	0	30	*	0
Deposit Interest	*	*	*	*	*	*	*	*	0.01
Loan Interest	*	*	*	*	*	*	*	*	0.03
Purchase National Bonds (NB)	0	0	0	0	0	0	0	*	*
Redeem NB	0	0	0	0	0	0	0	*	*
Accept NB by CB	*	*	*	*	*	*	0	0	*
Redeem NB from CB	*	*	*	*	*	*	0	0	*
Loans from Bank	0	300	100	300	300	0	*	*	1000
Redeem to Bank	0	0	0	0	0	0	*	*	0
Deposits in Bank	0	0	0	0	0	0	*	*	0
Withdraws from Bank	0	0	0	0	0	0	*	*	0
Product Price per Unit	0.2	0.5	1	6.25	11	*	*	*	*
Capital Investment (Numbers)	2	3	2	4	8	2	0	*	*
Sales of Products (Quantity)	770	580	420	14	21	*	*	*	*
Employment Numbers	70	70	60	30	65	330	35	*	*
Total Wages	90	90	80	30	90	430	50	*	*

TABLE 7.2 Institutional Parameters

	Adoption of a Polity: Institutional Parameters									
	1	2	3	4	5	6	7	8	9	10
Subsidy for Half of the Capital Investment	FALS	TRUE	TRUE	TRUE	TRUE	TRUE	FALSE	FALSE	FALSE	FALSE
Subsidy for the Deficit Makers	FALSE	TRUE	TRUE	TRUE	TRUE	TRUE	TRUE	TRUE	TRUE	TRUE
Subsidy for Half of the House Investment	FALSE	FALSE	FALSE	FALSE	TRUE	TRUE	TRUE	TRUE	TRUE	TRUE
Issue National Bond Guarantee of Central Bank	FALSE	FALSE	FALSE	FALSE	FALSE	FALSE	TRUE	TRUE	TRUE	TRUE

Table 7.3 Capital Investments

		Decision Making of Capital Investment								
Term	Machine available	Agri.	Flour	Bakely	Steel	Machine	House-hold	Gov.	Sum	Stock of Machine
1	26	2	4	1	7	11	1	0	26	0
2	35	2	5	4	10	14	0	0	35	0
3	44	2	5	6	10	19	2	0	44	0
4	55	2	10	8	10	23	1	0	54	1
5	62	3	15	10	10	22	2	0	62	0
6	75	2	15	17	12	25	3	1	75	0
7	81	4	17	15	10	28	4	3	81	0
8	81	5	21	13	16	24	2	0	81	0
9	88	5	22	13	10	33	3	2	88	0
10	97	3	30	15	13	32	3	0	96	1

Table 7.4 Management and Political Decisions

	Management & Political Decisions of Agents									
Term	1	2	3	4	5	6	7	8	9	10
Income Tax Rate	0.1	0.05	0.05	0.05	0.02	0.02	0.05	0.05	0.05	0.05
Corporate Tax Rate	0.2	0.1	0.2	0.2	0.1	0.1	0.1	0.1	0.1	0.1
National Bobd Rate	0.01	0.01	0.01	0.01	0.01	0.01	0.01	0.01	0.01	0.01
Official Bank Rate	0.01	0.01	0.01	0.01	0.01	0.01	0.01	0.01	0.01	0.01
Deposit Interest	0.01	0.01	0.01	0.01	0.01	0.01	0.01	0.01	0.01	0.01
Loan Interest	0.03	0.03	0.03	0.03	0.03	0.03	0.03	0.03	0.03	0.03
Price Wheat	0.2	0.2	0.2	0.2	0.2	0.2	0.2	0.2	0.2	0.2
Price Flour	0.5	0.5	0.5	0.52	0.52	0.52	0.52	0.52	0.52	0.52
Price Bread	1	1	1	1	1	1	1.1	1.1	1.1	1.1
Price Steel	6.25	6.25	6.25	6.5	6.5	6.5	6.5	6.5	6.5	6.5
Price Machine	10	10	10	10	10	10	10	10	10	10
Wage(Agr.)/ Person	1.2	1.2	1.2	1.3	1.3	1.3	1.3	1.3	1.3	1.3
Wage(Flou.)/ Person	1.3	1.3	1.3	1.3	1.3	1.3	1.3	1.3	1.3	1.3
Wage(Bake.)/ Person	1.3	1.3	1.3	1.4	1.4	1.4	1.4	1.4	1.4	1.4
Wage(Steel)/ Person	1	1.1	1.2	1.3	1.5	1.5	1.5	1.5	1.5	1.5
Wage(Mach.)/ Person	1.3	1.3	1.4	1.6	1.6	1.6	1.6	1.6	1.6	1.6
Wage(Gov.)/ Person	1.4	1.4	1.4	1.4	1.5	1.5	1.5	1.5	1.5	1.5

7.3.3 Human Gaming of a Bird's Eye View Model

We show the results of economic development over ten terms. The following are results of gaming simulation by a bird's eye view model across the terms under the parameters of the previous tables. The following Figs. 7.1, 7.2, 7.3, 7.4 and 7.5 show the results. Figure 7.1 shows the numbers of residents per house. At the initial term 27 persons are living in a house on average. At the last term living standard does not become good. Figure 7.2 shows GDP (real) and food consumption (BRU) per person. Figures 7.3 and 7.4 denote the change of price index, cash in the Government and acceptance of national bond by Central Bank respectively. Figure 7.5 shows the product stock of steel (STU) and machine (MAU). The acceptance of national bond by Central Bank works well in this case.

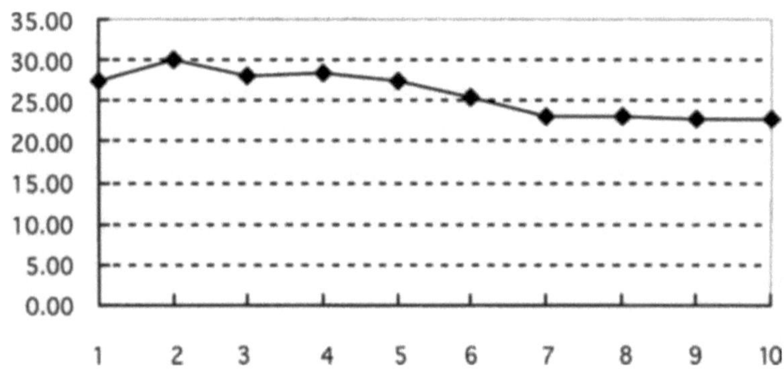

FIG. 7.1 The Number of Residents per House

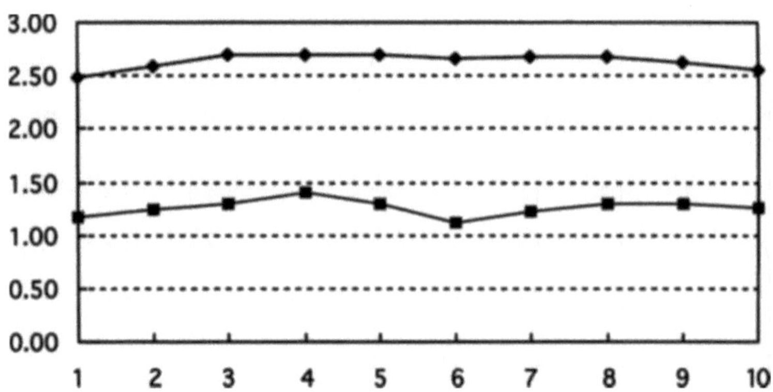

FIG. 7.2 GDP (Real) per Person and Food Consumption (BRU) per Person

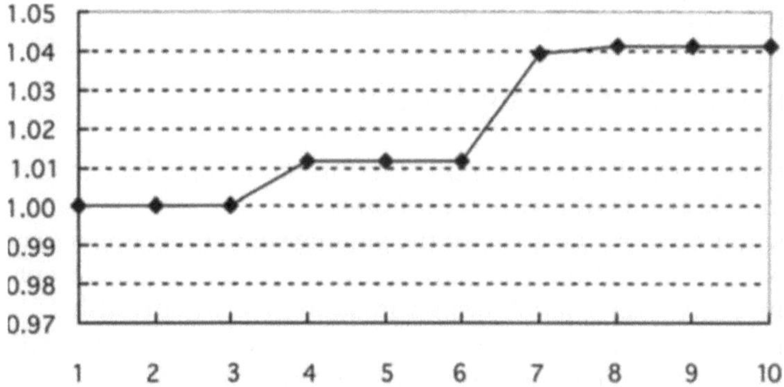

FIG. 7.3 Price Index

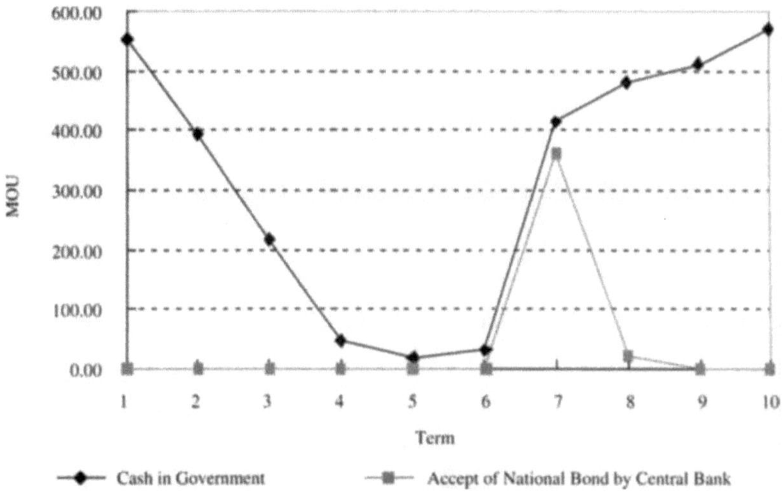

FIG. 7.4 Cash in the Government and Issued National Bonds Accepted by the Central Bank

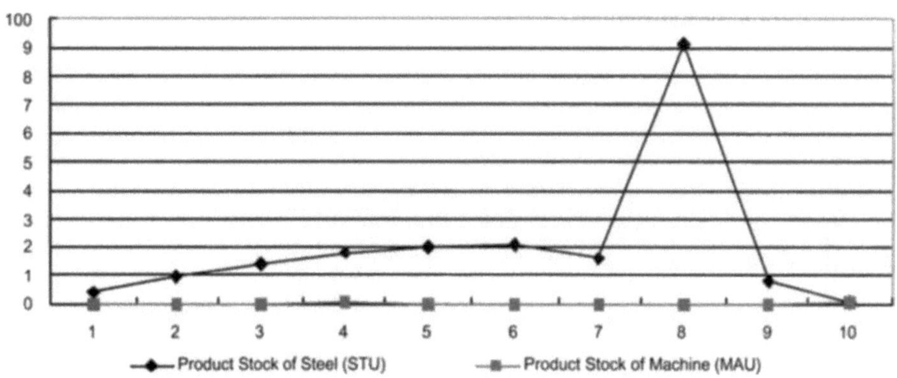

FIG. 7.5 Product Stocks of Steel and Machines

7.4 Multi Player's Gaming Model

The last gaming is a multi player's model. In this model nine human players play as nine economic agents of a virtual economy.

In this section we show the result of the gaming simulation as macro economic data graphs. The simulation was played by the under graduate students of Saitama University, over 7 fiscal terms and supported by spreadsheet software, designed using exchange algebra. In this scenario exports were permitted from term 3 using the domestic price and imports from term 7.

Figures 7.6 and 7.7 show consumer consumption and GDP and total imports and exports of this example respectively. Figure 7.8 denotes the change of invest-

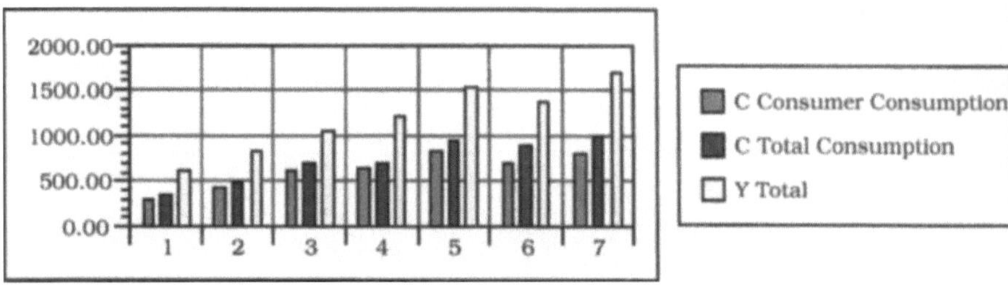

FIG. 7.6 Consumption and Gross Domestic Product (GDP)

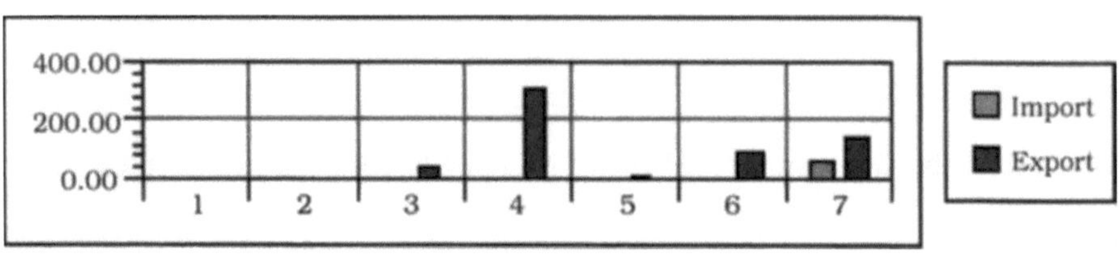

FIG. 7.7 Total Imports and Exports

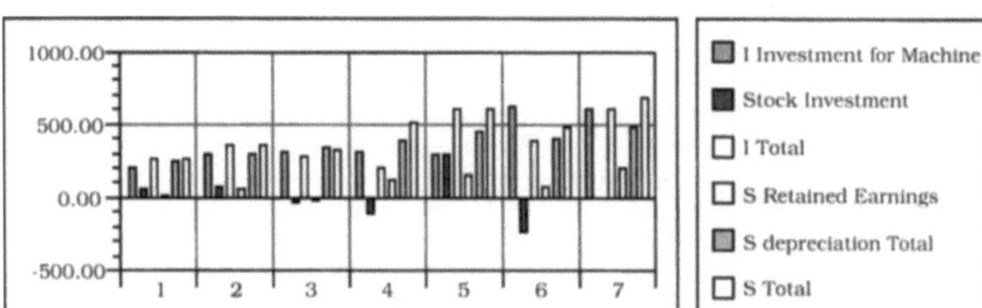

FIG. 7.8 Investment and Savings

ment and savings. Figure 7.9 shows the change of population (/100), GDP and food consumption per person. Figure 7.10 denotes the change of the corporation tax and income tax rate. Figures 7.11 and 7.12 show wages per person and central bank note balance. The government reduced the corporation tax rate on the 2nd term. The corporation tax rate rises afterwards and then drops because of the recession.

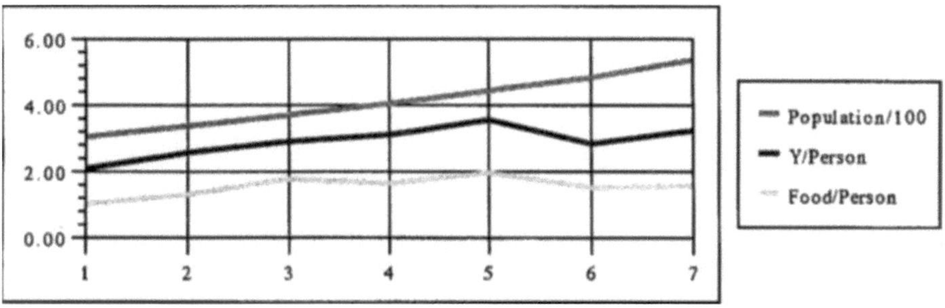

FIG. 7.9 Population and Food Consumption and National Income

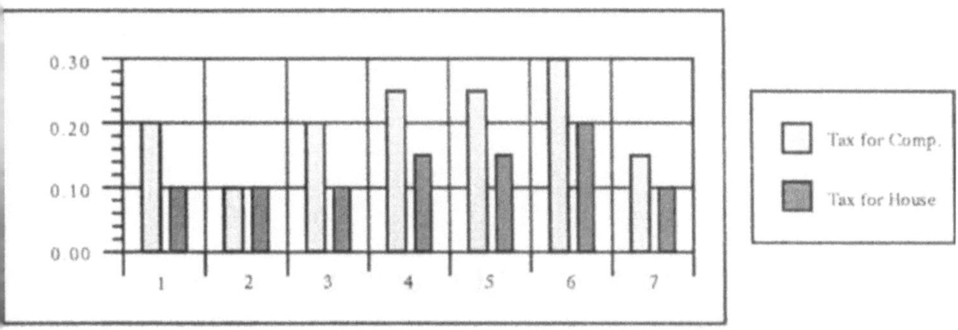

FIG. 7.10 Tax Rate Information

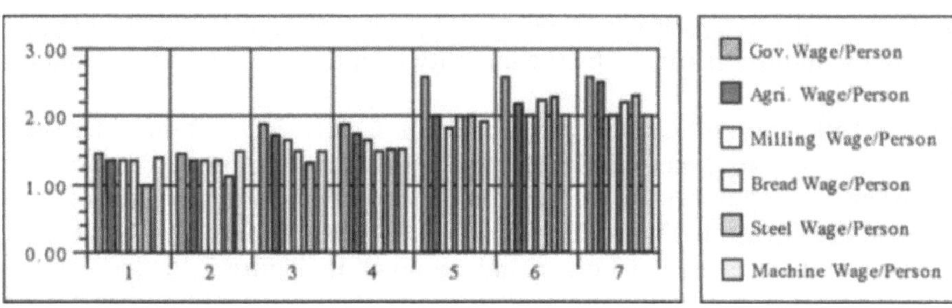

FIG. 7.11 Wages per Person

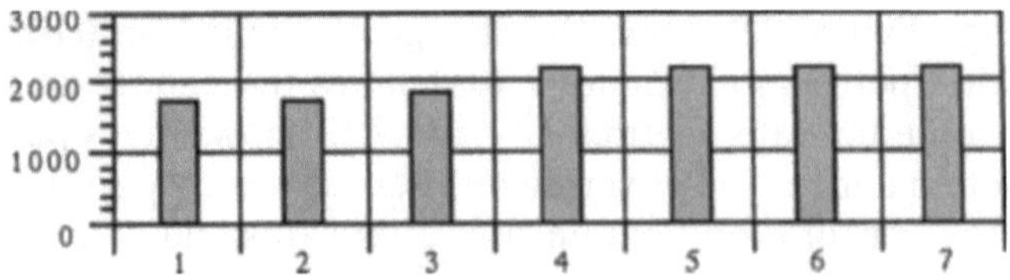

FIG. 7.12 Central Bank Note Balance

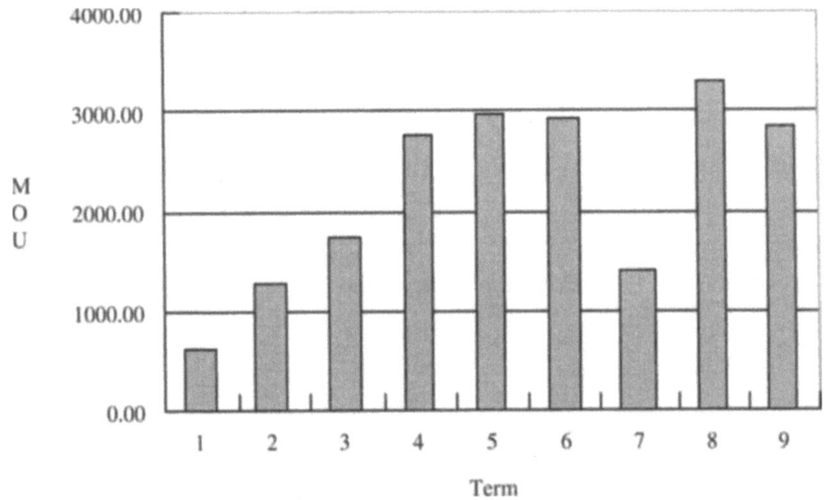

FIG. 7.13 GDP

7.5 The Failure and Success of Economic Growth by Gaming

(1) A Case of Failure: Crash

We show an example of the failure of economic growth in a gaming simulation by multi human players clearly showing a typical type failure of coordination among all the industries caused by the inability to anticipate the future demands of the economy.

The steel industry anticipated a demand for facility renewal and increased stock levels. The machinery industry also expected an update demand for machines and raised prices. As a result, demand declined, and the price of steel slumped at the 7th term. Figure 7.13 shows GDP (real). Figure 7.14 shows stock investment and investment in plant and equipment. Figure 7.15 denotes the change of prices of products. Figures 7.16 shows food consumption per person measured by BRU.

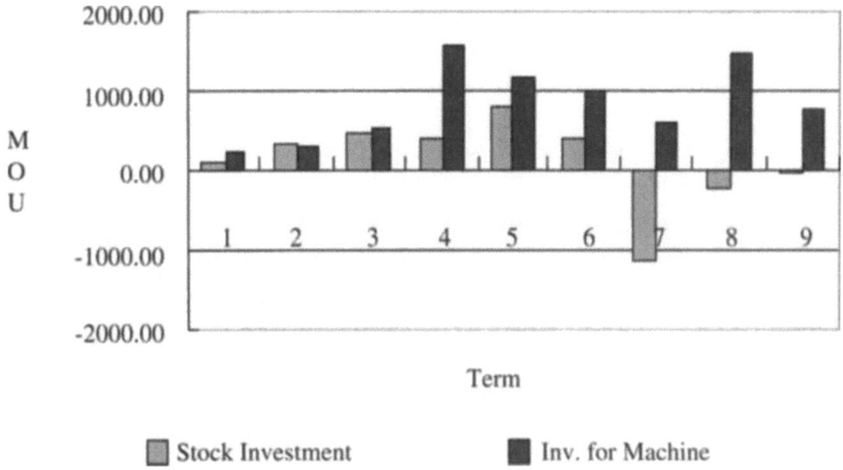

FIG. 7.14 Investment

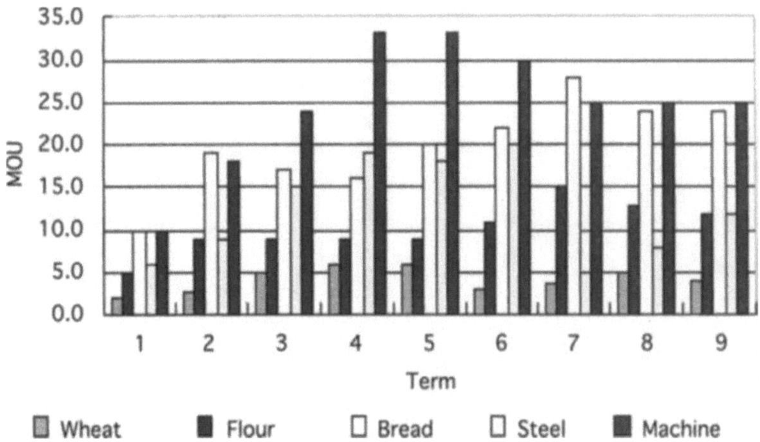

FIG. 7.15 Price

The social index in this economy is also very poor and food consumption is the minimum needed for survival.

(2) A Successful Case

In this case the Paasche Index shows price inflation exceeding 50%, but nominal GDP is very high and real GDP is growing strongly. Economic growth is successful. Figure 7.17 denotes the change of nominal and read GDP. Figure 7.18 denotes Paasche index. Figure 7.19 shows food consumption per person meas-

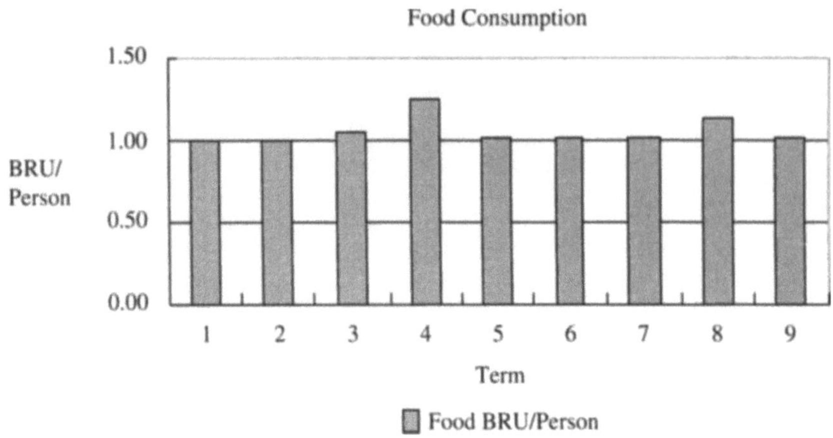

FIG. 7.16 Food Consumption per Person

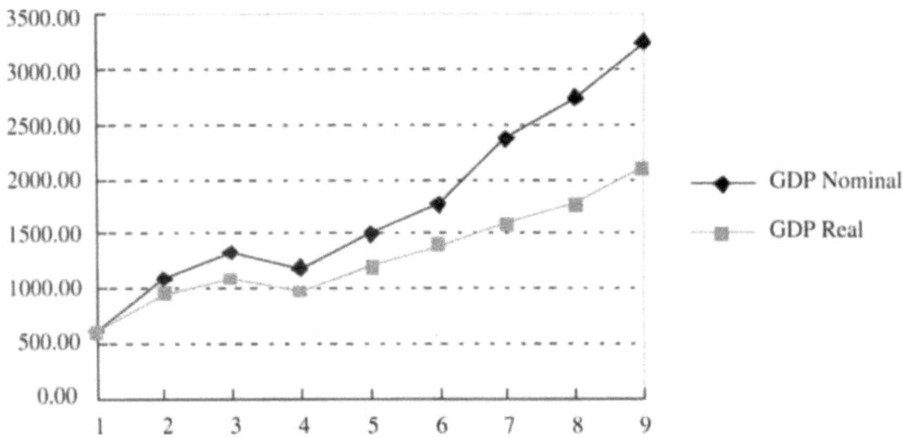

FIG. 7.17 Nominal and Real GDP

ured by BRU. After economic development, the social index improved, food consumption went up to 1.7 BRU per person and the number of people per house also improved.

Even in such a simple economy, we have demonstrated the extremes of failure and success.

Now, by using ABM, we will consider this economy, but first we will outline the work needed to compare the analysis of our virtual economy with an existing model.

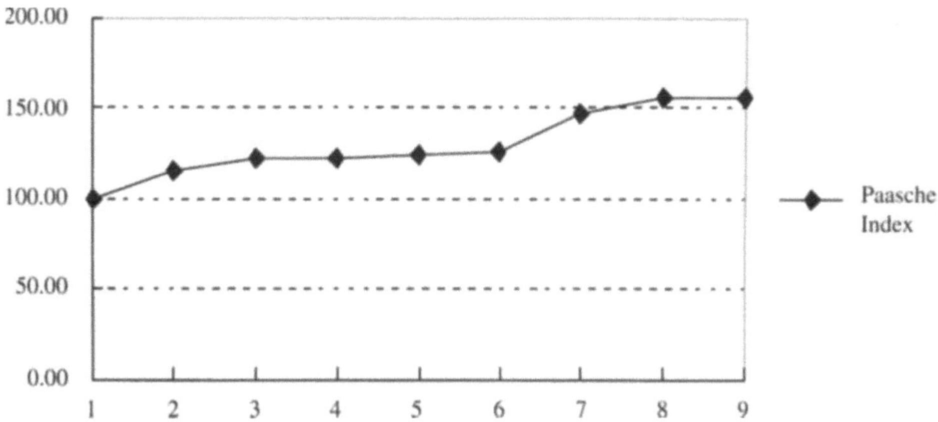

FIG. 7.18 Paasche Index

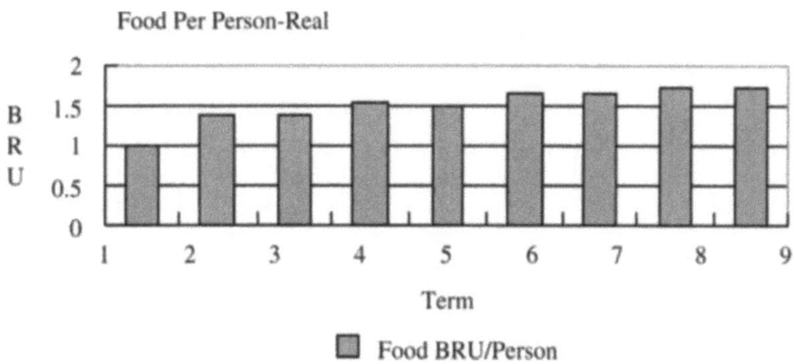

FIG. 7.19 Food Consumption per Person

7.6 A Multiplier Analysis of a Virtual Economy

We can apply multiplier analysis to the results of this virtual economy gaming. We can extract multiplier from our gaming data and then we can derive a linear regression model as follows. Figure 7.21 shows multiplier calculated from an observation of gaming. Figure 7.22 shows theoretical anticipated growth ratio of GDP by multiplier analysis and experimental data of gaming. A theoretical anticipated value is often fitted into an observed value after the experiment. We have to escape from this type of modeling method. CGE (Computational General Equilibrium Model) gives another modeling approach, but CGE or AGE are both too static for our purpose. Of course a forward-looking expectation model of econometrics and dynamic CGE model are now under construction as a protect belt of the classical macro and micro theories. Our research

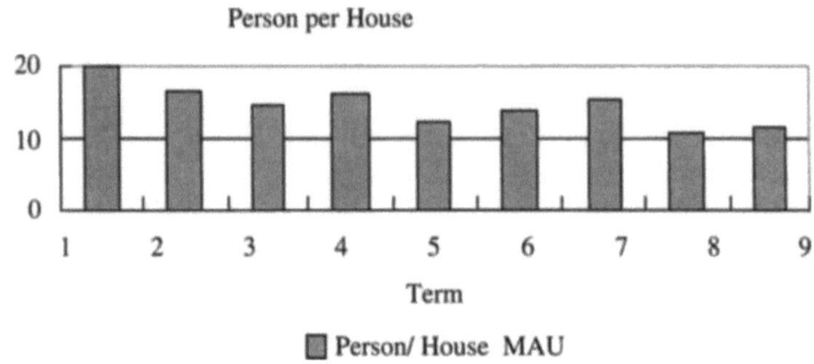

FIG. 7.20 Number of Residents per House

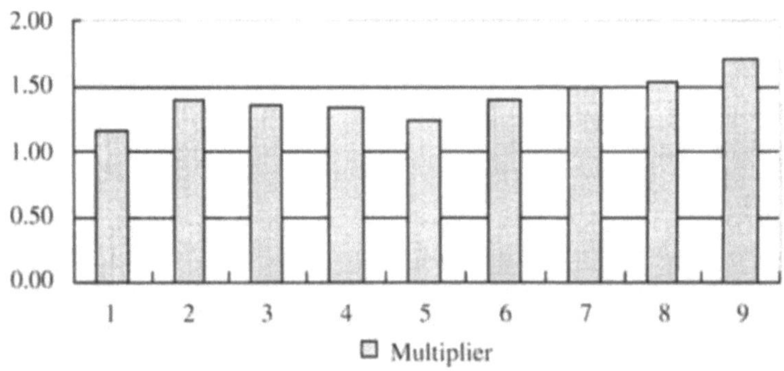

FIG. 7.21 Multiplier

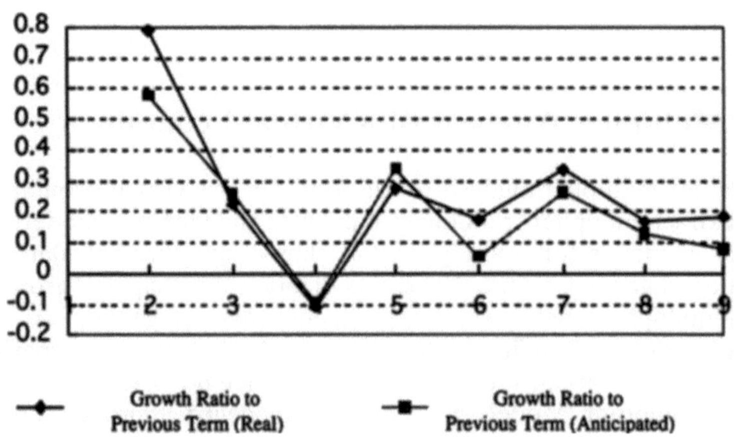

FIG. 7.22 Theoretical Anticipated Growth Ratio of GDP by Multiplier Theory and Experimental Value by Gaming

direction is different. We need to develop a dynamic theory of an agent based bottom up economy and its simulation method.

7.7 Agent Based Simulation of Bird's Eye View Model

Virtual Economy GA Simulation (VEGAS) is introduced for analyzing a bird's eye view model [Deguchi, 2003a]. We would like to optimize a path of economic growth across a certain period. For this purpose we define chromosome and gene as follows. Each period of economic growth becomes a gene locus on a chromosome. A gene is expressed by a vector. This vector represents a set of investment allocation ratios among the economic agents during the period. Elements of each vector consist of the investment allocation ratio of each agent such as "Agriculture", "Flour-milling Industry", "Bread Industry", "Steel Industry", "Machinery Industry", "Government" and "Household". Two points crossover and roulette tournament selection with an elite preservation strategy are used in our algorithm.

A chromosome represents a series of an allocation ratio of investment between several periods. This chromosome codes a phase of growth. We formulate an evaluation function to maximize the quantity of food consumption at the end of a term with minimum consumption during terms. At first we evaluated a chromosome by this evaluation function and later we introduced another evaluation function.

Figure 7.23 shows the number of residents per house by GA simulation, which is a typical social index in this economy. At the initial term 30 persons are living in a house on average and with the rise in the standard of living at the last term

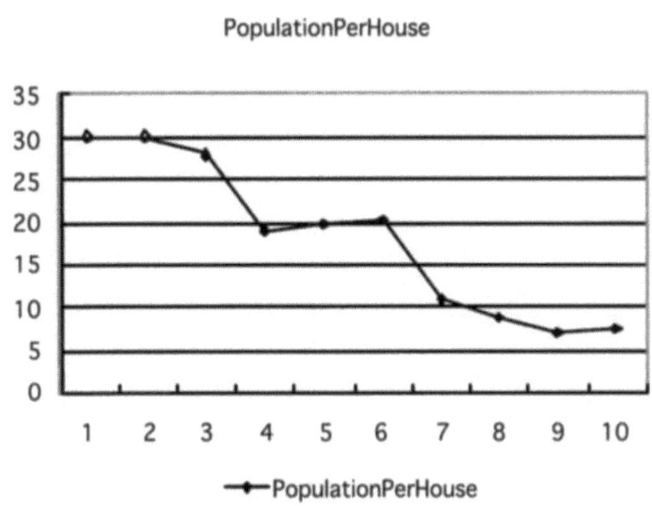

FIG. 7.23 The Number of Residents per House

this has fallen to 7.5 persons. The improvement process is very good. Figure 7.24 shows Food Consumption (BRU) per Person and real GDP per person. The growth path is better than in the previous human player's case. Our purpose in this economy is to attain food consumption of 2 BRU per person and in this case we have attained our aim.

Figure 7.25 shows the change of the investment allocation ratio among such economic agents as firms in the industrial sectors, households and government. The graph shows the investment characteristics of the effective growth path we have attained by using a genetic algorithm. Figure 7.26 shows the quantities of machine and steel sold in each term and Figure 7.27 shows the quantities of wheat, flour and bread, which were sold in each term.

7.8 Long Term Growth Path

7.8.1 Maximizing Consumption Utility Case

It is more difficult to find a long-term optimum or effective growth path in our model, because we have assumed a rapid increase of population such as $pop(t + 1) = 1.1 \times pop(t)$. This is a world described by TR.MALTHUS. We have

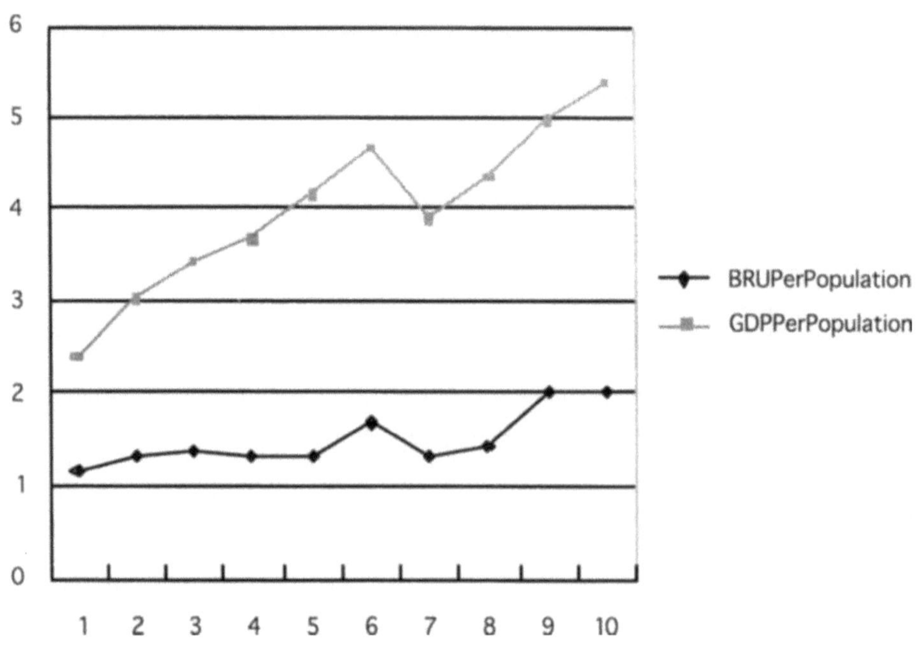

FIG. 7.24 GDP (Real) per Person and Food Consumption (BRU) per Person by GA Simulation

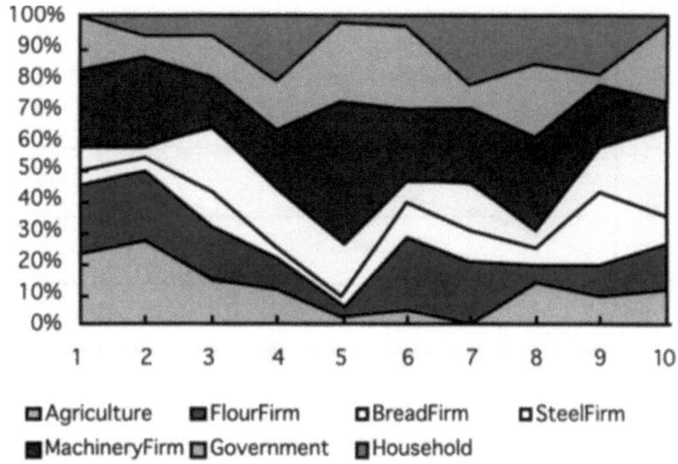

FIG. 7.25 Allocation Ratio of Investment Among Firms, Household and Government

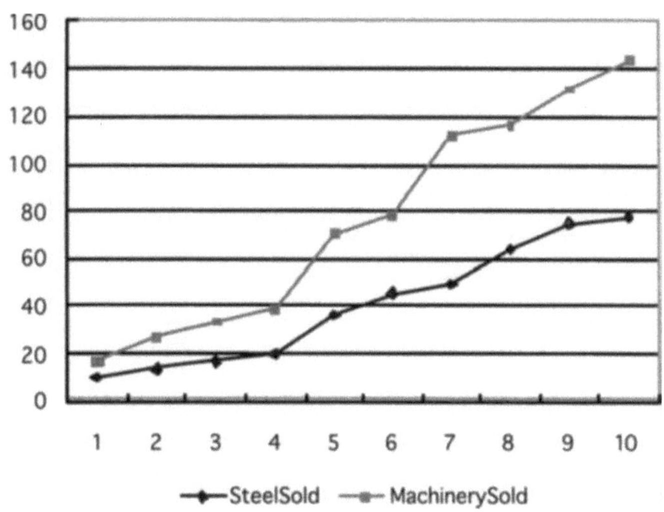

FIG. 7.26 Steel and Machines Sold in Each Term

tried to find a sustainable growth path in 20 terms with a utility, which maximizes the last consumption with minimum consumption at each term.

Figures 7.28 and 7.29 show a successful case of a long-term growth path with a rapid increase of population. Figures 7.30 and 7.31 show the increase in the number of products, sold during each term. In the last 4 terms there was an increasing tendency to change rapidly. This is a typical turnpike property under a maximizing consumption utility. These graphs also show a balanced growth rate of products except in the last few terms.

FIG. 7.27 Wheat, Flour and Bread Sold in Each Term

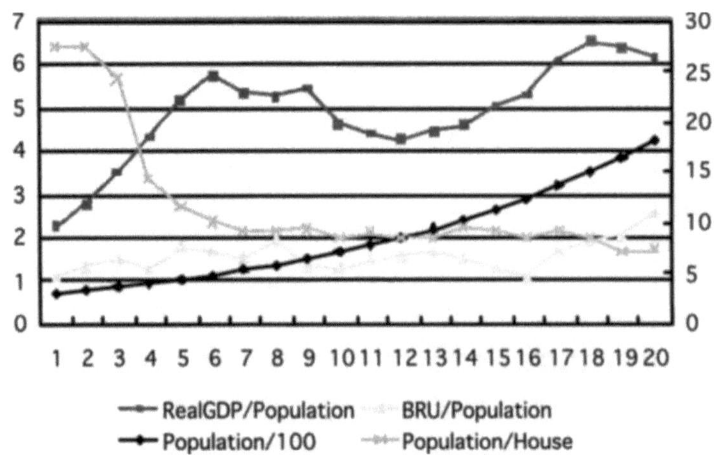

FIG. 7.28 GDP (Real) per Person, Food Consumption (BRU) per Person, Population (/100) and the Numbers of Residents per House

7.8.2 Maximizing Investment Utility Case

Figure 7.32 shows the growth path over 20 terms under a utility that maximizes both the last and the average investment. This utility makes it possible to attain better consumption as a result. Figure 7.33 shows the allocation ratio of investment in this case.

Figures 7.34 and 7.35 show the increase of the products sold in each term. In this case there is also a steady tendency towards an increasing number of products caused by maximizing the investment utility.

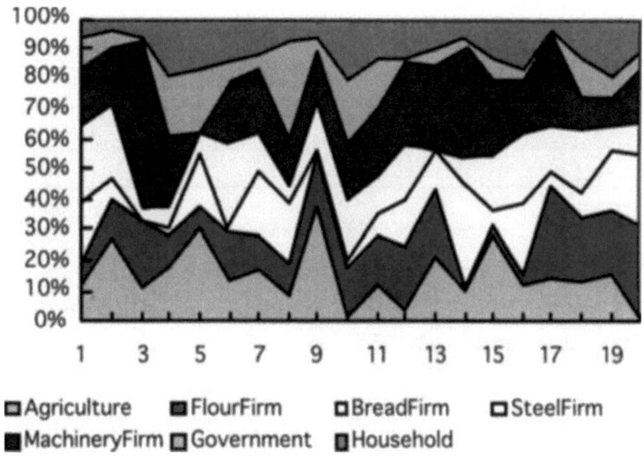

FIG. 7.29 Allocation Ratio of Investment over 20 Terms

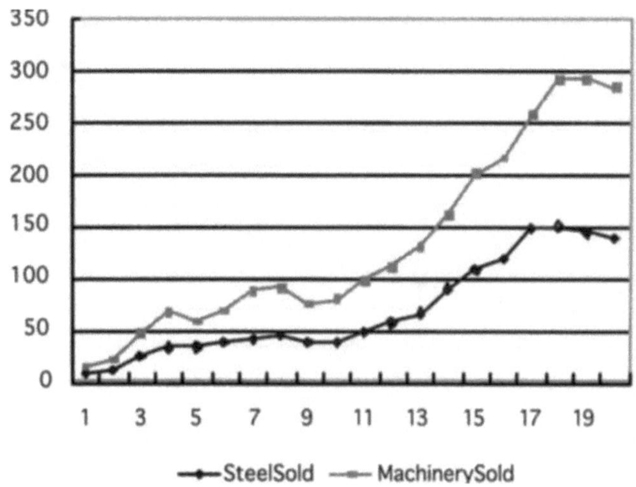

FIG. 7.30 Steel and Machines Sold in Each Term

Even if the growth path is balanced it does not mean that economic decisions such as the investment allocation ratio are steady. Actually our results showed that the investment allocation ratio is not a smooth and gradual path but a bumpy and winding road. It is quite difficult to distinguish the difference between the decision patterns of Figs. 7.29 and 7.33.

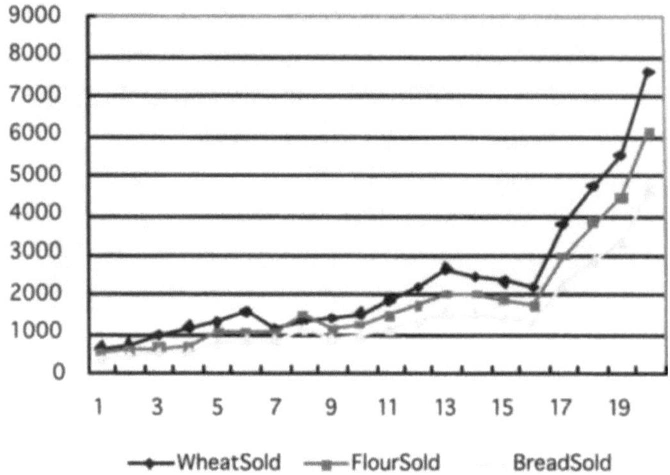

FIG. 7.31 Wheat, Flour and Bread Sold in Each Term

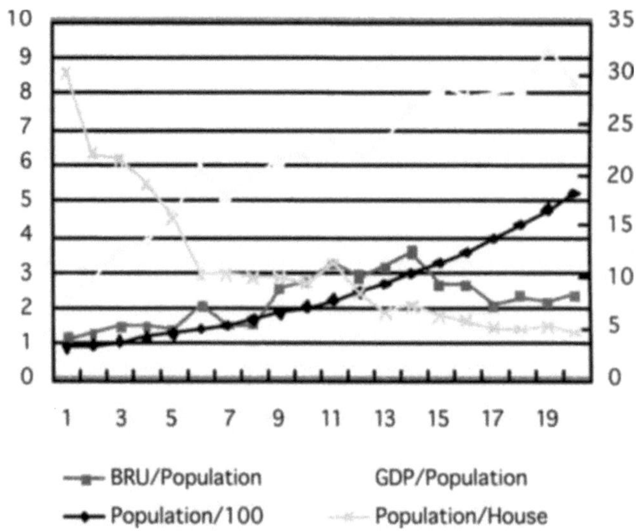

FIG. 7.32 GDP (Real) per Person and Food Consumption (BRU) per Person During 20 Terms

7.9 Conclusion

We investigated an agent based gaming simulation model of a small national economy. The model differs from the usual macroeconomics model as we have given an agent-based bottom up state description by exchange algebra and have also had to add a bottom up, multi agent decision-making mechanism.

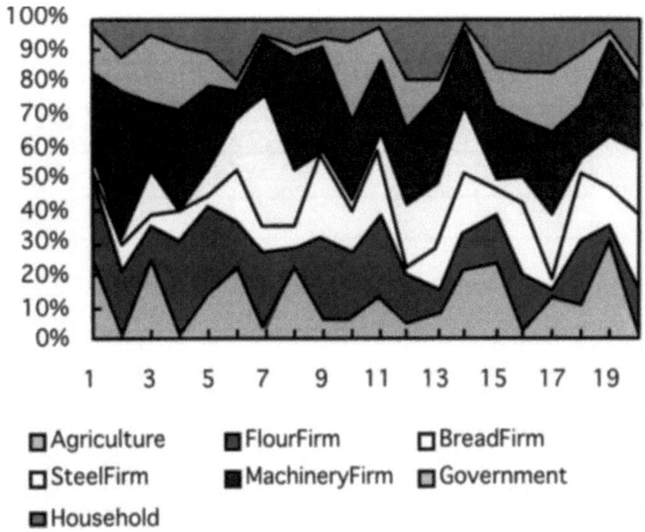

FIG. 7.33 Allocation Ratio of Investment During 20 Terms

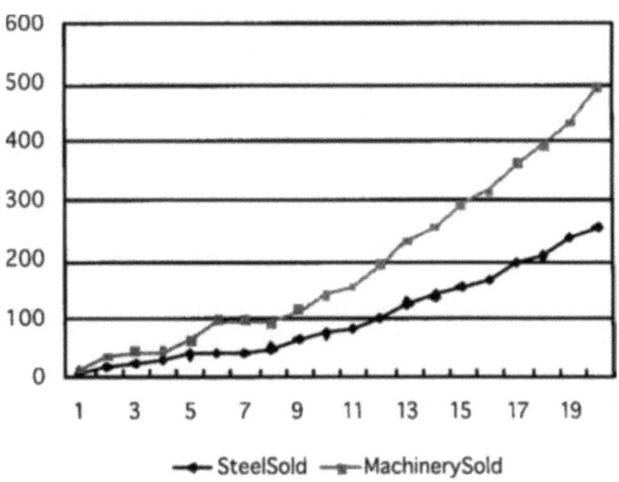

FIG. 7.34 Machines that Were Sold in Each Term

As a first step for our research program, we want to express and design the institutional and structural varieties of a real economy in the agent based models.

In this chapter we understood a national economy as a multi agent gaming simulation. The design of the gaming depends on our abstract algebra called exchange algebra and by using algebraic formulation, we can express not only the variety of economic states but also the variety of economic structures in the gaming model. We have investigated bird's eye view model by both gaming and

FIG. 7.35 Wheat, Flour and Bread Sold in Each Term

GA simulation. By using GA simulation we have analyzed dynamic growth path under complex boundary conditions and aims. Even in this simple model we find that the complexity of turn pike path and its management.

The decision-making activity of each economic agent changes through the process of mutual learning and commitment in the gaming. In this research stage it is difficult to make a suitable model of these types of learning and decision-making processes, all of which are observed not only among the game players but also in the real economy.

The players mutually refer to other player's decision-making actions and learn from each other. The game gives the boundary condition for the communication of economic activities and can be used as the means for a bounded communication of economic activities. From this virtual economy formulated by exchange algebra, players can learn an enourmous amount about economic structure and their effect on economic behaviors.

If we provide a more realistic and larger scale virtual economy model on a computer network, played by many actual business actors with the help of software agents, then virtual economy gaming would help the process of coordination, conflict resolution and self-organization of a real economy through a process of bounded communication.

Thus we would like to extend our model to include many agents and the differing models of other countries. Gaming simulation for mutual understanding is quite different from simulation for giving a prediction for rational decision-making. Our gaming simulation model gives a basic analysis for this purpose.

For our next research step we shall construct a large simulation depending on agent based modeling. This includes a research program for the reconstruction of the next SNA standard using our algebra and agent based simulation.

Chapter 8
Network Oriented Industrial Structures

Aim of This Chapter

In this chapter, we analyze the way that the "NETWORK" transforms the industrial structure. To accomplish this, we must first analyze the network concept and reassess the network industry as a platform industry.

Then we look at the changes in the structure of the computer industry; in particular, the software industry so as to analyze these changes from the point of view of network formation within the industry. It is argued that within the network, a knowledge network, or "open society" that transcends barriers between individual companies, has been constructed, and has been producing profits from the "improvement and advancement" of the software technology.

Finally, we pay special attention to the high industrial agglomeration of small and medium enterprises so as to analyze their network-advantages, and then, examine the possible steps for this high industrial agglomeration to become a knowledge-oriented transformation of the industry itself (note 1).

8.1 What Is a Network?

8.1.1 A Network as a Platform

The term "network" now attracts a great deal of management attention but when the word is used, the meaning is still vague. On the one hand, the term "network" is a concept, which has been derived from the fields of telecommunication, utilities, transport, and computers. On the other hand, it denotes the dynamic relations among companies or social actors. The two concepts, which have different origins, are now overlapping and with the development of new information technology forming a new synergetic concept.

The primary meaning of network is a physical channel (hard network) through which the service contents such as electricity, gas or water are distributed. Hard networks distribute and supply the service content in a single direction: from a large service supplier (server) to numerous terminal service users (clients). The organization that installs the physical network is normally the same organization that supplies the service within it.

With telecommunications networks on the other hand, the enterprise does not provide the communication service content as part of its basic point-point connection service. In the same way, even with value-added information services using telephone lines, the Dial 900 Information Service, for example, the provider of this information service and the provider of the communication network are mutually independent enterprises. We distinguish networks in the latter sense from the former by calling them network platforms.

When the term "network" is used, we often confuse the service delivered through a network platform with the service, which provides the platform itself. It is necessary to distinguish the provider of a network platform as the channel from the provider of the service, which is delivered on the platform (server). It is also necessary to distinguish the service user (client) from the platform and from the server on it.

In a traffic network such as a road transport system, the network platform is the road. It is normally provided as part of the public infrastructure. However, the network services, which use the roads, are provided competitively by large numbers of transport companies.

The usage of the platform and the services on the platform are stratified. In the case of Dial 900 for example, the dial 900 equipment (server) provides a service using the telephone circuits as its basic platform. At the same time, many information providers offer programs using the equipment as another higher rank platform. This type of stratification of services (functions) is, as we will explain later, clearly seen in the computer field.

In a case in which large scale physical facilities are required to provide a network platform service, economically, the externality of the network (economic efficiency of system scale) becomes an important issue. Here, however, by the "externality of the network" we mean the convenience, which the scale of the network provides to the network users. Because it also signifies the competitive advantage of enterprises that have been able to form a large scale network, in this sense we also refer to it as the economic efficiency of the system scale [Hayashi, 1989; Deguchi, 1994a].

In other words, when network platforms are in competition with one another, each network enjoys competitive price and customer attraction as its share of the market increases. This means that network services such as gas, water, and electricity suppliers are permitted a form of monopoly, and in compensation for this monopoly they are subjected to restrictions in light of their public nature.

The merits enjoyed in such cases by the supplier of the service which is delivered on the network—a supplier who is integrated with the enterprise that builds the network—is based upon the economic efficiency of the system scale, which in turn depends upon the size of the network. Essentially, the provider of a network platform and the provider of services using the platform do not have to be the same.

The history of the development of networks themselves has a record of cases in which the potential providers of a service on a network were narrowed down and these providers were allocated the task of building the network platform. When it is possible to separate services on the platform from the platform itself, and to guarantee some form of connection between the platforms provided by different enterprises, it will be possible for businesses to competitively establish the platforms and for other companies to compete to offer the services that will use the platforms.

In a case where it is possible to separate the platform provider from the service provider so that the network platform can be offered to the service provider at a suitable price, the issue of the externality of the network changes. Particularly, in a case where each service provider can transversely utilize a number of platforms, the economic efficiency of the scale of the network will have little effect, and cost and functional competitiveness will come to the fore. Then it is also difficult to obtain strategic merits by vertically integrating and enclosing a network platform and the services provided on it.

Take expressway networks for example. There is no exclusivity between platforms (a plurality of expressway networks are connected anywhere), and even if we assume that there are a number of expressway networks owned and operated by different organizations, the strategy forming connections with one of these networks would be meaningless for the operators of truck fleets and other providers of transportation services. Rather, depending on the cross-over of the networks, the network platforms acquire the character of public resources in the sense that they are used in common for competition at the service level.

When a network platform is a type of common-use environment or public resource which can be used by a variety of service providers, the economic efficiency of system scale will become a benefit for users of the network platform, but it will not necessarily provide competitive advantage to the platform providers. Of course, in the gas, water and other classical hard network industries, it was technologically difficult to even consider splitting the platform provider from the service provider on the platform. Roads on the other hand, can be handled this way. When a network platform and network services can be divided as they typically are on today's new information networks, it is possible for a number of services to compete in some way on shared network platforms.

8.1.2 The Concept of a Network in Sociology, Economics, and Business Administration

In contrast to the network concept found in these network industries, a network as used by recent management science, means a type of intermediate structure or a dynamic relationship between a core firm and related firms [Mitchell, 1969]. In these cases, the term, "network" refers to a relation which is weaker than static role structures in an organization, but which is stronger than a relation in a market on which only prices play as the integration variable.

However, even when the network concept of this type of dynamic relationship between (compound) nuclei is the object of consideration, the point of view varies. The point of view will, for example, vary depending upon whether the issue raised is the weak relationships which lead to the break down of the internal structure of existing corporate organizations through measures such as forming manufacturing process networks by out-sourcing production processes as proposed in the symbolic management theory of R. Reich [Reich, 1991], or whether the issue is the weak relationships which are created by the introduction of relations such as that with subcontractors in the market for intermediate goods: networks formed between parent companies and their subsidiaries for example. Another view stresses networking between creative people within management organizations, yet another places weight on on-the-spot information, and emphasizes the role of networking as a type of organization that reorganizes relationships. Others are developing a theory of information culture in which small creative groups are active as they form knowledge work networks based upon the similar formation of information networks. These varied points of view are delicately blending to form the present day versions of network organization theory, industrial theory, and a theory of civilization.

If networks such as the convenience store networks in the distribution industry that are connected to the end consumers and therefore able to serve their demand, can be linked directly to factories, thereby cutting out the wholesalers, the industrial structure will, in all likelihood, undergo sweeping changes.

When the terms "network organization" or "network within organizations" are used, they reveal a confused use of the many varied network concepts currently in use, and the prospects for deliberation on the issue are poor. At the very least, we have to categorize the many distinctive concepts of network found in industrial society, and analyze the problem to determine what sort of benefits each of these concepts of networks as weak relationships can provide for each type of organization.

8.2 Industrial Networks in the Software Industry

In the previous section, we presented a brief summary of the network concept found in the hard network industries and of network concepts postulated by the social sciences, management science in particular. These two concepts of a network, now linked permanently through the present revolution in information technology, are opening new possibilities. This trend is conspicuous in the information industry. This section is an examination of this issue.

8.2.1 Network Platforms in the Information Industry

We have already considered the concept of network as it applies to the hard network industries. It is possible to discuss information and communication networks in the same terms. The network platform used to transmit signals for an information provision system such as a television broadcasting station is considered the right to utilize electromagnetic waves. In contrast to this, the broadcasting equipment is the server, which, along with the party that provides the programs broadcast using the platform and the parties who receive the programs (clients) form the network service. With present day systems of this type, the party providing the network platform is integrated vertically with the parties who provide the programming that is broadcast on the platform. On top of that, the extent of these systems is controlled so they operate within nations or regions, primarily under government regulations.

It is, however, now possible for the network platform provider to be separate from the broadcast service provider on cable type broadcasting networks such as CATV. Regardless of the fact that it is possible for transmission systems themselves to be interconnected in this case, under present circumstances, the party establishing the network, the network platform service provider, is the party offering the broadcasting service for each region in accordance with government regulations. As stated above, however, the provider of a network platform is not necessarily the party, which, at the same time, provides the service delivered on the platform. It is also difficult to find grounds for a network separated by regional boundaries.

With a system such as CATV, one oriented towards communication using a transmission network permitting high volume advanced use, it is possible for various service providers including television broadcasters using the network to be separated from the party providing the network platform.

Computer based data communication networks are one more type of network. Information, in the form of data packets, which come in a variety of formats, travel back and forth on a network which provides channels for physical (electronic) signals. Accordingly, the achievement of economic efficiency of system scale on a network of this kind is significantly affected by the interconnectivity of the standards of the various protocols in the system: not simply by the physical connections of the network.

In recent years, we have seen a remarkable trend in the computer and household electronic industries. They have offered services on platforms provided by equipment and standards of some type, and the popularization of these platforms has brought the economic efficiency of scale to these systems.

In this sense we will expand the discussion from problems related to the relationship between the providers of services that use network platforms and the clients to include more general problems concerning the form of an industry which is able to use equipment, software, and standards of some kind as a platform, and to provide services of some kind on this network platform. We will call this industry the "platform industry." We will call the sector of this industry that provides the platforms themselves, the "platform provision industry." In contrast to this, we will call the industry which, premised on the existence of some form of platform, provides services for this platform the "platform service industry."

The hard networks described earlier, are classical platform industries, and as these platforms have grown, both the platform provision industry and platform service industry have enjoyed the economic efficiency of system scale. They coexist, sharing the added value. With a platform industry of this kind, when the platform provider must also be the party providing the service on the platform, or in other words when the platform provision industry and the platform service industry are vertically integrated, the further the platform spreads, the greater the economic efficiency of system scale and the larger the profits earned by the platform provider.

Even if the provider of the platform can be separated from the provider of the services on the platform, service provision is concentrated on platforms, which are large scale systems, and small scale platforms are at a disadvantage in competition with other platforms. Thus, as we saw in the competition between household electronics makers pushing the beta and VHS video formats, struggles surrounding platform standards and their popularization constitute a vital side of the business strategy in the platform provision industry.

Inversely, if a platform provider can provide a service for which there is great demand, or can link up with or dominate a platform service provision company that offers services that are in high demand, the platform provider can gain relative superiority in the platform provision industry. This is called the strategy of vertical integration on a platform.

Members of the platform provision industry grope for various kinds of cooperation and enclosure strategies of this kind, so that the technically superior platform does not necessarily survive. As argued by Katz and Shapiro, however, it is not so simple that one can state confidently that standards supported by suitable sponsors will survive [Katz, 1986]. At the same time as Nintendo's Family Computer provides a platform, it has achieved vertical integration by selecting those companies who provide services for the platform.

In this way, it provides an example of a strategy, which successfully achieved the benefit of the economic efficiency of system scale. However, even if the platform equipment or system provision industry successfully achieves vertical integration of its service, the industry will not necessarily enjoy large profits. The platform provision industry has promoted efforts to open standards or to create groups of various kinds based on low license charges because the spread of their standard is the key to its success. However, as the classic case of the IBM-PC indicates, there are cases in which the platform provision industry itself is so structured that it is difficult to obtain added value through excessive competition aimed at expanding system scale. Today it is said that software dominates hardware in the computer industry, and in the same way, it is possible for much of the added value to be earned by the parties providing services for the platforms (platform service industry).

Even in such cases, however, for the industry as a whole, the platform element serves as a type of public infrastructure (of course, this varies from the traditional sense of the term "public"), and makes a substantial contribution to the improvement of the productivity of overall industry.

The interesting thing about the present-day computer industry as a platform industry is the fact that while the companies in the industry are competing fiercely in both the hardware platforms and in operating systems and other software platforms, computer technologists have, by creating free software and IP interconnected information networks such as the Internet, voluntarily cooperated in the construction of beneficial platforms which can be shared by software technologists.

This characteristic of the industry can be seen in the process by which the UNIX operating system was created, the development of the TCP/IP protocol, and networks that use this protocol. Once such a technology is able to gain the overwhelming support of an organization of technologists, it acquires far greater strength in business competition than it would have if confined to a single company.

This means that within the computer industry, the achievement of compatibility between platforms at many levels is a critical issue. At the same time that the achievement of compatibility is an advanced technological issue, it is also one related to competition between companies and the distribution of added value among them, so this is not a simple issue. In cases where the platform provider and the party providing the services on the platform are identical or are vertically integrated, popularizing one's own platform in the course of competition with other platforms can result in monopolization of the service, and bring profits to the company providing the platform.

Providing a service on some type of platform is widely approved within the present day information communication industry, and these platforms are stratified to create an environment for service users. For example, the most basic platform is the computer hardware. Normally, the operating system forms the strata directly above the hardware, and the application software treats the operating system as its platform. It is permissible to see the windows environment as a platform lying between the operating system and the application software. In addition, computer networks themselves are a type of platform, on which services are provided through relationships between servers and clients.

The concept of platform is, in this sense, stratified. A situation in which a specified service (operating system or application software) is provided on a number of platforms so it is not dependent upon one specific platform can probably be described as a case of a service provided by crossing over a number of platforms. When this is done, we say that there is a crossover between competing platforms or that these platforms crossover one another.

These crossovers, like platforms, have a stratified structure. Examples of stratified crossovers include the ability to use the same operating system or windows environment on computers with differing hardware and cases where multimedia software can run on different operating systems. Crossovers between platforms are important in that they alleviate the disadvantages of closed off markets and technical rigidity caused by differences in standards in the platform industry and by vertical integration of services on platforms, and create environments conducive to the competitive development of superior technology.

Another way to describe a platform crossover is as something, which is stratified so that higher (more practical) functions are not restricted by the lower level functions, which make them possible.

The stratification which separates functions of this kind from restrictions on their realization have long been treated as the problem of free connectivity by those involved in computer network protocols. In more general terms, this issue has expanded from stratification of hardware, operating systems, windows environments, application software, and so on to include the question of stratification of functional objects. As a characteristic of systems found in all areas of the new knowledge industries, it is a noteworthy feature.

According to the arguments of Katz and Shapiro, in circumstances where there is a sponsor, and supplementary income is earned, standardization is possible even for low level technology. When platforms crossover, however, the user needs services on the platforms, and if it is possible to use such a service by crossing over a number of platforms, the externality of the network concerned with the platform itself will either not be realized or it will be weak.

If platform crossover becomes possible at different levels, the relative importance of the economic efficiency of scale of the system will be low for the platform provision industry. Rather, inexpensive platforms will spread throughout the industry as a consequence of competition between platforms.

If a fall in the price of platform use affects the expansion of the scale of all industries that use the platforms at the same time as the use of the platform contributes to more economically efficient productivity, it will help improve productivity in the economy as a whole. Unlike value added distribution trends caused by the economic efficiency of systems scale, this will have a significant effect on society as a whole.

8.2.2 Industrial Structure as Connections of Objects

Having already expanded this discussion from a consideration of network platforms in the hard network industries to encompass platform industries in general, we will now take the framework to an even more general level by introducing the concept of networks of connected functions in the industrial structure.

In order to discuss the merits of standardization in a broader sense than that used when considering how the platform industries are established to provide services on the platforms, we have to expand our consciousness of the problem. Here we will focus on the functional interconnections between parts in a product formed through the assembly of standardized parts as a network of functional connections in the formation of a manufactured product. If we focus on the interconnected character of the functions, the services in the platform industry and the interconnected character of the platforms are also networks as functional interconnections. In networks of mutual connections between parts in a manufactured product or of functional interconnections that provide services on a platform, the final products and services are considered to be the downstream element, and there exists a mutual relationship with demand which considers the platform, which is based on essential components and services, to be the upstream sector. All the manufactured products of industrial society are collections of functional elements.

A television set is a collectivity consisting of a frame, metal dies to attach the internal electronic components, a cathode-ray tube, power source circuits and various functional blocks of electronic circuitry; each electronic circuit block contains circuit boards to hold each block and on each circuit board, there are integrated circuits, capacitors, resisters, and other electronic components.

In a word, all manufactured products created in industrial society are made up of large or small collections of functional parts of this kind. These functional components are integrated by forming functionally linked networks between parts or between parts and the final product (to be precise, at this level, a branching structure composed of upper and lower levels), with the achievement of each specified functional component requiring the assembly of lower level functional components.

The interrelationships between functions of this kind are, needless to say, also formed in the sphere we refer to as the platform industries. The purest form of the interrelationships between functions in this case is seen in the computer world: the provision of software premised on specified hardware platforms such as the IBM PC. One characteristic of platform industries of this kind is that, as the sale of computers without software and CD players without recorded music indicates, the final consumer market for the platform itself is wide open, and the platform itself has value as a product. In general, with the stratification of components used to produce industrial products, individual components have value only in markets composed of manufacturers who use these individual parts to make higher level functional components or final manufactured products.

We will use the term "functional object" (or simply "object") to express the process of combining functional components of this kind. We will tentatively define objects as input output systems with specified information and operations related to this information as units. Objects, which are unrelated to the internal structure used to achieve its functions, are independent functional units that need only be observed externally within the function specification range. Any type of functional component is designed as a functional block that provides a specified output reaction in response to a specified mechanical or electrical input. These are also called objects. Functional objects of this kind can be found in various spheres: from parts of mechanical circuits with physical properties to electronic circuit components, and even software components. The integration of functional objects to create functional objects with more complex functions does, therefore, accord with the process of creation of the manufactured product.

The concept of object is itself, a concept developed from an awareness of problems related to the fact that the software industry creates software components and constructs complex software by combining reusable components. In the case of the software industry, the design of functional objects is done using specified software languages such as Java, and it is possible to realize it with hardware on which this language is implemented. One reason for changes now occurring in industrial structures, which are affecting the hardware side in a variety of industries, has been a process in which functional objects inseparably linked to this gradually become intellectual objects and software objects. To put it another way, it is now possible for such intellectual objects as design to be treated as independent matters because it is now possible to separate manufactured product design from manufacturing.

In addition, the conversion of functional objects to software has transcended borders between companies and permits the accumulation of know-how concerning the broader design of functional objects.

8.2.3 Knowledge Oriented Transformation in the Computer Industry

In the present-day software industry, the task of making application software operate on a specified hardware platform (target machine) is inseparable from the functional design of the software itself.

Even in the hardware industry, the standardization of many parts has progressed so that once a part has been standardized; it can be manufactured in many ways despite the fact that its use depends upon CPU design. Even the CPU, which is the heart of a computer, has been exposed to the interchangeable chip strategy, and there are no limits on the methods used to achieve the functions of the original developing company.

Under these circumstances, it is not easy to keep technical knowledge of functional objects within a single company. Furthermore, even consumers are conscious of all products from the computer industry as collections of assembled functional components; and this has altered the industrial structure creating leeway for the creation of new added value by companies, which make system integration their business.

The standards of manufactured products are themselves brands, and as long as a prescribed product quality is maintained, the method used to realize the product and the manufacturer have no brand name value.

In these circumstances, the computer industry has seen a trend to try and protect programs, CPU design and so on by strengthening patent and copyright laws, etc., while on the other hand, an understanding that, regarding programs in particular, the importance of sharing knowledge and know-how, and to mutually feed-back this information, is spreading widely among technical groups. Even corporate managers have now begun to realize that a technology kept within a particularly company is not a technology that will come into use throughout the market [Dell, 1998]. Members of technical knowledge networks composed of both power users and technologists now look for the value of a company in a framework which lets them discover its role as an organization serving as a nodal point formed to invest appropriate capital and carry out technical finishing touches to transform technological know-how into manufactured products of some kind.

This point of view is made possible by the fact that when a network transcending the boundaries companies place around technological knowledge is once created, it begins to function as a network that provides benefits by improving and encouraging the evolution of technological knowledge. When an industry reaches this stage, its individual corporations cease to be concentration points for information about production technology, and are no longer points where manufactured product knowledge is concentrated.

But simply because they are still companies where capital is concentrated, they can carry on concrete development of manufactured products by investing suitable amounts of capital in such enterprises, and in this sense, the importance of corporate organizations has definitely not declined. Rather, it is now possible that, with technological knowledge the common property of a collectivity of technologists in the broad sense which includes end users and power users, companies operating in this way will permit the formation of networks transcending corporate barriers, but which are composed of technologists employed by corporations, and can be counted upon to function as indispensable entities that permit knowledge networks directly linking technical knowledge to the varied requirements of ordinary people. Companies will coexist with broad external technological knowledge networks as an important strategy followed to gain the benefits of improvement and evolution.

Because the products of the computer industry are nothing more or less than equipment to support their own communication activities, the changes we define as the formation of networks transcending barriers between companies' technologies, which we have focused on, have just begun to appear and have been determined. In other industrial fields, even those where it is possible to separate the production and the design of products, and manufacturing is no longer capital intensive, there is nothing to prevent the spread of product design knowledge beyond corporate boundaries and the formation of networks comprised not only of technologists, but users of their products and services. However, we believe that in many industries, this process is irreversible. More than anything else, the division between production (actual) and function design increases the possibility that many individuals and groups will contribute to product design in the field in question. In addition, once an open network concerned with functional design of a product is formed, the merits of improvement and evolutionary progress will begin to act upon this product, and the network will eventually acquire technological improvement capabilities transcending those of an individual company.

When this happens, coexistence with this network and growing through its use will, for the corporation, be a strategy providing competitive advantages. The computer industry is undergoing this change exactly as we have described it. It is unlikely that any computer language designers have not been influenced by Richard Stallman's Free Software Foundation (FSF) GPL, or its products. Even technologists that have not been influenced by "gcc" and other FSF free software, cannot ignore the power of EMACS as a development environment. LINUX is another example of the power of GPL.

As a result of knowledge oriented transformation of the industry, the work of the standardization of product categories and the creation of technological platforms in many products are in progress. When, under these circumstances, a network providing common access to knowledge of a product category to members outside a single corporation comes into existence, its assessments and discussions will have an influence on standardization that cannot be ignored, and this influence will govern market trends.

It is possible that technological trends will be far more sensitive to the effect of de facto improvement and evolutionary progress in standards achieved on such networks than to the influence of such markets and efforts to coordinate standards by governments participating in international bodies formed to coordinate interests between nations. In fact, the TCP/IP protocol and the various services it provides as part of interconnections in information networks were tested and discussed on the Internet itself, and their improvement and evolutionary progress continues using RFC (Request for Comments) documents. In contrast, the international standard OSI has not come into wide use.

A corporation is, above all, a point where the people, things, money, and information required to develop, manufacture, and sell products are concentrated.

In the manufacturing industries in particular, the important role of a company is to serve as a place where technological raw input and capital meet under appropriate conditions to mature the technological input to create manufactured products. There are now doubts that the organization of an extremely large company is an appropriate setting to transform technology into manufactured products. Let us look at a couple of examples from the computer industry.

The RISC chip is based on IBM technology, but it was SUN that brought it to life, and GUI technology developed by Xerox was endowed with overwhelming technological influence with the appearance of the Macintosh by Apple Computer. In industries where technical change is rapid, and there is a high degree of uncertainty about which technology will grow, there are many examples of cases in which a company, successful in the past, adheres to its past success, and is unable to take full advantage of new technological opportunities. Overgrown corporations tend to gradually solidify so they cannot take advantage of their human resources: the innovative ability of their personnel.

Turning to technological development on the other hand, as knowledge becomes the dominant form of technology, there is a tendency for the capital required for such development to shift away from investment on equipment to investment in human resources. This does not happen in industries requiring concentrated investment in equipment: the pharmaceutical and semiconductor industries for example, but these trends are conspicuous in the computer industry. For this reason, an important strategy for a company to follow in order to increase added value and to gain competitive advantage is to take advantage of the concentrated technology, which can be described as "external knowledge resources," held by technologists linked together by a network such as the Internet.

8.3 Networks of Medium and Small Enterprises

8.3.1 An Industrial Agglomeration Network

So that we can deal with an industrial agglomeration consisting primarily of small and medium enterprises as a network, we will examine a network functionally, and consider it as functional connections between various industrial functions. In order to do so, we have adopted an analysis of industrial agglomeration proposed by Watanabe [Watanabe, 1992]. Watanabe has stated, "The more diverse the forms of machine industry-related factories and the more factories established within a range which permits face to face transactions, the greater the level of industrial agglomeration." Generally, the reasons why a company assigns work to subcontractors can be broken down as follows. To begin with, there are cases in which a subcontractor is commissioned to perform a process not performed within the company so that the equipment necessary for the process need not be acquired. This is the commonest form of Japanese style subcontract processing activity. The classic example is that of a large corporation surrounding its major plants with cooperating subcontractors to form company towns. In other cases, large factories have work performed by subcontractors in other regions or other countries where the work costs less than it would to carry out the process itself. In recent years, so called fabrication-less companies, or companies which do not produce physical goods, have built their own design centers, but have aggressively commissioned the production work to low-cost outside companies so that they need not invest any capital in production facilities of their own.

A second approach is for a company which has the technology and equipment to perform a particular process to supplement for the insufficient capacity of this equipment by having outside firms carry out that portion of the processing that exceeds its own capacity.

In other cases a company will commission outside companies to perform tasks requiring specialized technologies: hardening, plating, heat treating, surface processing, machining, press work, a milling, extrusion molding, external and internal polishing, etc. which the company cannot do itself.

As specific examples, Watanabe cites and discusses two machine industry concentrations: one in the Keihin District near Tokyo Prefecture and another in the Okaya and Suwa regions of Nagano Prefecture. According to his analysis, the machine process order reception and production type corporations in the Keihin District share the following characteristics:

(1) Orders are received for a broad range of jobs subject to wide variations, but not large volume jobs.
(2) There are many advanced specialized processing technology companies.
(3) An important source of many orders is medium and small companies producing their own products.
(4) Orders are received from all parts of the Kanto area, including small and extremely small companies.

As another characteristic of this concentration, Watanabe points out that an intimate mutual order network has been formed linked by a physical distribution network, which in Watanabe's own words, "is in a range that permits companies to ask for evening shipments and expect early morning deliveries." He points out that while on the one hand, this distribution network has been improved to adapt to the breakdown of the old structure which formed enterprise company towns, it has cultivated an industrial agglomeration capable of taking advantage of its advanced specialized technological strength to respond to demand fluctuations.

In the Okaya and Suwa region on the other hand, companies receive diverse orders, but in many cases, they receive large orders. By diversity, he is referring to diverse production volumes and diverse quality requirements, and to technological diversity. The latter case can be described as qualitative diversity related to production technology (note 2).

The role of the medium and small enterprises is to meet the requirements presented by this diversity. To handle technological diversity in particular, the key is to provide an environment that permits the integration of advanced technologies. This is a characteristic of the industrial agglomeration he found in the Keihin District. Medium and small companies in regions such as the Okaya and Suwa Region vary in that while in-company specialization is advancing, other aspects are marked by interdependence seen in the use of outside commissions. At the same time as the completion of interdependent networks lets the local industry deal with diverse changes in the content of orders, it allows for mutual distribution of the profits.

It is characterized by close cooperative relationships accompanied by fierce competition, which destroys companies that lack technological capabilities. In this sense, the situation could be described as an industrial agglomeration based upon the technological specialization of individual companies, with a network that provides the benefits of improvement and evolutionary progress to a prescribed degree established without the coordination function that a specific parent company would provide [Deguchi, 1992a,b].

In a competitive—coexistence relationship characterized by both competition and mutual ordering between companies, a structure takes shape in which companies accumulate supplementary distributed specialized technology, which is then evaluated. This inter-company relationship is, however, fundamentally based on face-to-face relationships, and the formation of a network based on human relationships dependent upon direct communications.

These small and medium enterprises themselves formed these networks through mutual outside order activities, but this is a principal characteristic of the concentration of small and medium enterprises in the Keihin District, Accordingly in the Keihin District, most ordering is for products which in Watanabe's words, "are unit products such as prototypes etc. which vary widely, but require advanced technology." This indicates that the network within the concentration of small and medium enterprises in the Keihin District constitutes a network which has common access to distributed advanced knowledge and technology related to manufacturing.

In an industrial agglomeration possessing advanced technology of this type, one category of network externality (economic efficiency of system scale) is at work providing one of the benefits of a network: the fact that if one company in the network commissions work from another, the company receiving the commission will, as part of this technologically linked concentration of companies, be able to handle the commission creditably. Within the industrial agglomeration, the merits of improvement and evolutionary progress are enjoyed thanks to the mutual competition, technological refinement, and information exchanges taking place within the industrial agglomeration.

The existing subcontractor relationships found in Japan formed environments in which the benefit of improvement, cost reduction, and quality improvement were enjoyed as part of the ongoing relationship between the parent company and its subcontractors: relationships which created relatively closed networks coordinated by the parent companies. The limits to the pursuit of benefits through the formation of exclusionist networks of this kind are, however, gradually becoming pronounced amidst changes such as the internationalization of the industrial structure, the recent knowledge oriented transformation of the industry and the appearance of companies that do not perform manufacturing work themselves. Our purpose here is to point out that networks formed by industrial agglomeration possessing advanced technology have the potential to replace this older type of network.

As possible developments in the industrial agglomeration of medium and small subcontract firms, we will now look at two possibilities observed among structural changes in the industrial system in recent years. The first is the possibility of further development of industrial agglomeration among small and medium enterprises, which will undertake more work as the amount of manufacturing work commissioned to outside companies increases. In this way, they will serve as the hub of manufacturing for various companies.

This can be broadly categorized into one of two cases: cases in industrial fields where standardization of manufacturing has progressed or the technology has matured so the principal factor behind the selection of a company to undertake an outside commission is cost, and cases of outside orders which include the development, prototype manufacture, etc. within industrial concentrations which have the capacity to manufacture a wide range of more advanced, higher added-value products. The former is a pattern seen in its most typical form in the horizontal deployment of production centers to take advantage of the reduced cost of overseas labor.

The second possibility is the appearance of concentrations of small and medium enterprises as groups which, having established networks comprised of a group of technologists or of organizations of technologists and even some users in order to share, accumulate, improve, and advance product knowledge—a type of industrial organization we call an open society—act as autonomous organizations in aggressive efforts to build up and make use of their respective networks.

This will necessitate the formation of a network, which enables the groups to accumulate and to exchange advanced technology information. This second possibility is one that is still in the development stage. But when we study approaches to outside commissions which are not dependent solely on cost considerations, we find that the possibility that industrial agglomerations will develop as networks that pursue the merits of evolutionary progress and improvements is one way that an industrial agglomeration with advanced technology might develop as a location for production and manufacturing activities including the prototype production and even design-in by fabrication-less companies. For this possibility to be realized, the use of an information network is indispensable.

8.3.2 Providing an Industrial Agglomeration Network with Information Capabilities

At this time, industrial agglomeration networks are networks based on face-to-face interaction, and they not only do not share and accumulate technological information in electronic form, they do not even have satisfactory systems for handling order information as electronic information. It is absolutely essential that these advanced industrial agglomeration networks be provided with information network capabilities, if the industrial concentration networks are to be opened and if they are to mature into networks which will provide the economic efficiency of system scale and the merits of evolutionary progress and improvement.

We have touched upon the possibility that the addition of information networks to the industrial agglomeration networks will create open societies of advanced industrial technology. These networks have the potential for transformation into networks with even greater possibilities if they mature into industrial agglomeration networks that will produce manufactured products to satisfy final consumer demand, rather than simply being industrial agglomeration networks of subcontractors.

Another strategy manufacturers can implement to deal with technical uncertainty is to take advantage of industrial concentration networks. With technology changing as rapidly as it is today, it is not only impossible for a company to keep all of its new and innovative advanced production technology to itself, it is becoming a dangerous choice considering the cost-effectiveness aspect of the issue. In addition, knowledge oriented transformation of the industry is accelerating this tendency. With regards to production, it is increasingly often the case that it is not necessarily the most beneficial policy to adopt a strategy of investing in production equipment and hoarding production technology within one's company, in an effort to achieve a cost-saving economy of scale.

The product cycle is accelerating and diversifying, and in fields marked by high-speed technological progress, excessive investment in production equipment, which increases a company's depreciation burden, is a dangerous choice threatening its survival. In such cases, outside subcontractors are necessary. Companies in an industry that has fully standardized its production and uses parts that can be manufactured with relatively simple technology will likely choose the cost-first strategy of overseas expansion. However, if a company requires advanced production technology, and seeks cooperation at the product development stage, it is possible that an industrial agglomeration network with advance technology will be an important place to have its outside work performed. Let us again sum up the relationship between a technologically oriented industrial agglomeration and the fact that such concentrations are now regional concentrations. The benefits of a regional concentration are: (1) intimate human relationships, (2) shared access to information, and (3) convenient physical distribution. One model of this is the Keihin District.

One important possible approach to transforming one of these regional concentrations into a structure, which is independent of geographical proximity, is the use of the Internet. If we discover that the benefits of concentration include, "If one goes there, he will be helped in some way, or will be able to find some kind of clue or a solution." In this statement, "there" refers not to a specific location; "there" could be an information network. In this case, we interpret the merits of concentration as our ability to obtain appropriate answers to diverse unsystematic technical problems, by going there. The key is that we know that we will obtain a suitable answer by asking someone. Information letting us know "who knows what" is one form of information that is most often exchanged on the Internet.

In fact, advanced industrial agglomerations in the software industry can be said to exist on the Internet. In the computer industry, shared possession of the Internet has created a type of software industry concentration.

We wish to consider the following question. Is it possible for the structure of the Internet formed by the computer industry—what we can refer to as a virtual industrial concentration—to be duplicated by industrial agglomerations formed by small and medium enterprises that produce components, carry out machining, press work, a milling, extrusion molding, external and internal polishing or heat processing, etc?

8.4 The Merits of a Network and Networks of Medium and Small Companies

8.4.1 Inequalities that Reveal the Merits of Networks

When we consider changes now occurring in the industrial structure from the point of view of economic efficiency and merits for corporations, it is possible to sum up the situation by pointing out two kinds of structural change in progress. These are indicated by the following inequalities.

(1) Merits of out-sourcing production processes
> Merits of enclosing production processes
(2) Merits of sharing information and knowledge
> Merits of enclosing information and knowledge

The first inequality, which is already considered a merit of companies, which perform no manufacturing in-house, is a constituent factor of the economic efficiency of linkages, and the aspect, which now receives the greatest attention.

It is possible to interpret this inequality from one of two contrasting points of view. The first approach looks for the merits of production process out-sourcing in cost reductions. This is stressed by symbolic management theory and it is also a strategy seen in the horizontal division of labor characteristic of recent Japanese manufacturing activities in Asia. According to proponents of the other point of view, out-sourcing is beneficial because it allows the creation of networks, which permit the improvement, and evolutionary progress of technology, which in turn helps industry to adapt to technological change, refine superior technology, and improve its ability to reduce costs.

Frequently, a large company monopolizes its innovative new technology, but the failure of its bureaucratic management prevents its cultivation, and the company fails to effectively reduce costs and improve quality. The second of the two points of view described above proposes that companies obtain merits by solving such problems through continuous information exchange and the distribution of needs-oriented information on a network, which includes outside companies, so that they participate in a non-stop process of dynamic technological improvement and evolutionary progress.

The first approach is the pursuit of static financial benefits; the second can be described as the dynamic pursuit of structural merits. We forecast that the objective of forming industrial networks will swing from the pursuit of the benefit of cost savings to the pursuit of the merits of improvement and evolution.

At any rate, we are aware that the merits of out-sourcing now constitute an important strategy for corporate survival. A corporate policy of not keeping production equipment enclosed within its boundaries would have been an unthinkable strategy in the business environment of the high-speed growth era, when the economy of scale and the economy of range were established. In industrial sectors where vast sums are invested in production process equipment to establish the economy of scale, this inequality has still not appeared. The environmental conditions required for the realization of these inequalities include standardization and knowledge oriented transformation of the industry, and rapid technological change.

8.4.2 Two Possibilities of Knowledge Oriented Transformation of the Industry and the Centralization and Distribution of Knowledge

Changes in the structure of industry such as industry-wide knowledge sharing and the new ability to separate the design and production departments permit corporations to try a number of strategies. In environments where technology is changing rapidly and investment is extremely dangerous in particular, a company can now adopt a strategy of out-sourcing all manufacturing processes and specialize purely as a design center. This is one of the most conspicuous changes in corporate behavior to appear in recent years. It signifies a decision by a corporation to abandon the practice of enclosing its production equipment within the boundaries of its own system through heavy capital investment, and the pursuit of the economic efficiency of production scale and range (diversification). Take Nintendo Co. for example, this classic example of a fabrication-less corporation put together its Family computer development strategy while avoiding production-related risk.

Now that knowledge and design are the source of corporations' value added, each company will probably attempt to enclose them within its boundaries and protect them through patents and other means. However, we are focusing on another potential outcome of knowledge oriented transformation of the industry. It is possible that within industries sharing knowledge, knowledge will spread beyond corporate boundaries to form a knowledge network, which is even linked to consumers. In the computer industry, this sort of knowledge oriented transformation of the industry; the movement of this knowledge beyond the boundaries of individual corporations, and the formation of networks are actually passing from the possibility stage to become a reality.

When a corporation has separated the production of products from their design, and even its design activities can no longer be monopolized within its boundaries, what is the significance of this transformation for the company? There would seem to be little incentive for a company, which has isolated its production and design activities, to take its product knowledge and design know-how, which do in fact constitute its lifeline, and take active measures to provide it to outsiders. Nevertheless, if we admit that such a policy is possible, we ought to be able find some reasons for implementing it.

We conceive of one layer of the structural changes in industrial society as something within industry-wide knowledge sharing, the distribution of this knowledge, and the formation of networks. This concept does, of course, depend upon a number of conditions. It is still not that clear whether the industry-wide knowledge sharing that is now in progress results in this knowledge eventually taking the form of distributed knowledge within networks that transcend corporate boundaries or whether it will be concentrated within corporations.

If, even though production and design can be separated, companies have to invest vast sums of money to create the knowledge required for product design, then this knowledge will probably accumulate inside companies, which will, as in the past, serve as places to concentrate capital. Because biotechnology and pharmaceutical companies have to invest vast sums in research and development in order to create knowledge, these industries possess a knowledge concentration type of industrial structure. In the pharmaceutical industry, it is possible to separate medicines production from medicines development, so that the production of medicines, which have been designed, can be completely independent of development know-how. For this reason, companies, which cheaply produce medicines whose patent has run out, have appeared to fill a specific industry niche. Put this way, an important strategy for individual companies in the pharmaceutical industry is to protect their knowledge by means of patents and similar measures. This is because considerable financial investment and specialized knowledge are required to carry out product design at biotechnology firms and in the pharmaceutical industry. But even in fields such as medicine development and biotechnology, if, for example, it is someday possible to design the molecular structure of a medicine using an inexpensive work station and the knowledge required to do so is shared through a computer network, it will then be possible for product knowledge to escape beyond company boundaries, be distributed, and produce many specialists working outside of corporations.

8.4.3 Open Technology Society

Once a product knowledge network of this kind is formed as an open network, it develops greater power than that of an individual corporation. Corporations complete plans enabling them to benefit through coexistence with these networks. We wish to use the term "open technology society" to define a network created in this way by a new group of technologists, including users of the technology or products in question. This network is, however, not a network formed by people's interest in a single technology or problem. The network is formed by sub-networks concerning many related technologies, which cross over and compete landscape of evaluation one another. The network changes its form continuously as it evolves through countless information exchanges and experiments.

At the outset, the formation of an open technology society may be recognized to be a threat to the existence of corporations. Even in the computer industry, the computer network culture as an example of the culture of the west coast of the U.S. was for a long time, a heresy in the eyes of managers, if not to technologists. They finally revised this view after gaining an understanding and adopting a positive view of the way in which companies such as SUN and APPLE were able to take advantage of the rapid introduction of down sizing to develop new successful products using RISC chips and GUI: large companies such as IBM and Xerox developed superior innovative technologies but were unable to do anything with them. The technologists who supported these companies simultaneously created the Internet culture of the open society.

In order to remain in operation, a company has to compete successfully. In order to do so, it searches for various strategies for making profits. In this sense, it is natural for companies to adopt a strategy of maintaining a monopoly on its own production technology or turning to the legal system to protect its product design know-how and the designs themselves. However, such a strategy has to change to adapt to changed circumstances and a new environment.

We have referred to the fact that industry-wide knowledge sharing and diversification of needs are affecting many industries, endangering management strategies that depend upon economy of scale and call for a policy of keeping production technology and production equipment within corporate boundaries. At the same time, if an environment is created within which a corporation must allow its product design knowledge, which is considered the lifeline of a corporation, to be evaluated and improved on a new network of the kind we call the open technology society, the corporation will have to adapt to this situation.

Industrial systems of this kind can be viewed as one form that benefits society as a whole, because technology on a network improves and evolves in shorter cycles than it would if it were restricted to a single company, and this ultimately contributes to the accumulation of public wealth and the improvement of productivity. Products of all kinds are evaluated and improved or else developed on a network which includes both technologists and specialists outside the companies which produce them, and product users, which means that product design can be linked directly to a wide selection of diverse sources from all sectors of society.

An industrial system of this kind permits the continuous change and development of products in response to social diversity thanks to the non-stop improvement of products and rapid feedback of proposals from users, and increases the likelihood that products will appear that are suited to the requirements of each region and of society as a whole.

The industrial system can also change whenever necessary to meet the diverse needs of the public, reducing the chance of a slump in business activity because of a mismatch between demand and supply.

Notes

(1) This chapter was originally written as Japanese paper in 1994 just after the acceptable use policy on the Internet is released [Deguchi, 1994a,b]. The trend described in this chapter is now on progress. Now some examples are out of dates thus we replaced them. The mathematical analysis of the platform industry by using social learning dynamics and some other contemporary cases are given in the next chapter.

(2) But in these days (2003) situation has changed drastically and small lot of order are now popular in this area.

Chapter 9
Learning Dynamics in Platform Externality

9.1 Platform Concept

The purpose of this chapter is primarily to discuss the concept of a platform within a framework common to a broad range of areas from culture to industry. The concept of platform goods plays an important role in digital society, but it is also an extremely important concept in considering the structure and function of our socioeconomic system itself.

Subject to a fundamental service that is provided by a platform, the concept of platform represents a place that realizes supply and demand of various application services on that platform. This concept has applications in a number of industrial areas, such as the network concept in industry as we have shown in the previous chapter.

However, this concept of platform also has much broader applicability. This discussion will begin with a look at the platform concept in industry and the economy and then consider the potential of the platform concept as a design concept for a new socioeconomic system.

Apart from the platform concept, we also consider the concepts of network and network externality. The discussion of so-called network externality looks at the question of interaction between consumers and network providers. In this chapter, however, we analyze interactions among three groups, namely providers of goods called platform goods, providers of service goods or application goods that can be provided based on this platform, and their consumers. This concept of platform is a concept used in the computer industry, but adopted by the author as a concept to represent industrial structures as is shown in the previous chapter [Deguchi, 1994a,b].

Platform goods are likely to experience lock-in. This lock-in is the result of a dynamic change in the effects of platform goods, which is caused by a share of the platform goods or the accumulation of application goods on the platform goods. These effects have been discussed as an externality of platform that is an extension of classical network externality. The relationship between platform and service goods that is based on the platform includes the relationship between the transmission network of electric power and electric power supply by power generation and the relationship between PC and operation systems (OS). Further, they include the relationship between operating systems, regarded as higher platform goods, and application software, the relationship between video players and video software, and the relationship between CDs and music software. Platform goods include hierarchical relations such as the relationship between PCs and OS or OS and application software. Figures 9.1 and 9.2 illustrate these relationships. Table 9.1 summarizes the relationship of platform goods and the provision of service based on the platform goods and use of these services.

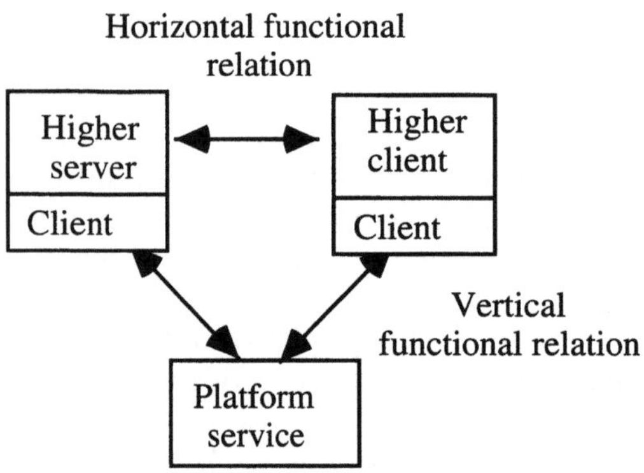

FIG. 9.1 Relationship Between Platform and Its Higher Client and Server

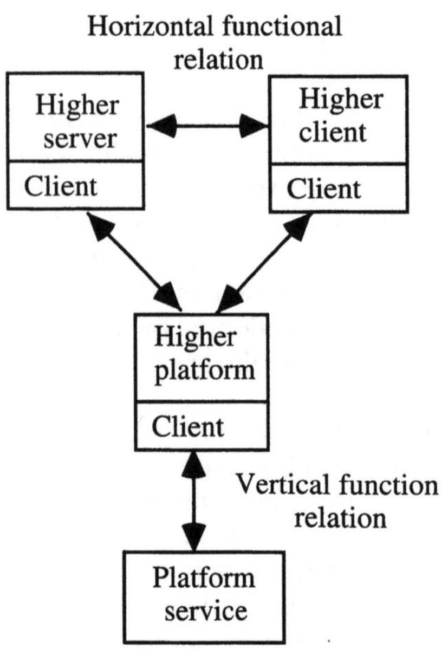

FIG. 9.2 Multilevel Structure of Platform

TABLE 9.1 Platform Goods and Client Server Relation on the Platform

Platform goods	Service goods	Client of the service
Provision of telephone lines	Telephone connection service	Telephone users
Electric power transmission system	Electrical supply service	Electric power users
Gas distribution system	Gas supply service	Gas users
CD player	CD music provision service	CD music listeners
Game machines	Game software	Game users
PC	OS	OS users
OS	Application software	Application software users
Private line service	Internet provider service	Internet users
Internet provider service	Servers on the Internet	clients on the servers

Goods in information society can often be divided into platform goods that enable higher-level services to be provided based on these goods (or services) and into the higher-level service goods themselves. In this case, most of these platform goods have little utility on their own, just like CD players without software to play. It is the additional value of service goods on the platforms that add value, even to their platforms. This relationship between goods generates a special interdependency of utilities. This hampers incentives and may create an environment of unfair competition.

This chapter conceptually studies cases that enable an analysis of the two dynamics: an alternative chosen by consumers of platform goods and an alternative chosen by providers of service goods. The analysis is made based on the results obtained by using an analysis method called social learning dynamics of decision-making that has been developed in the Chapters 2 and 3 and that is designed to find out what meanings these results carry in technological innovation. The chapter also analyzes what kind of institutionalization enables fair competition.

A lock-in is likely to occur in platform goods because the payoff of platform goods consumers depends on the payoff of service goods provided on the basis of these platform goods and because (1) the consumers' choice of platform goods and (2) service providers' choice of platform goods are linked to each other through their shares. This is stronger than the bandwagon effect and likely to create a market monopoly as a result. Platform goods may not be so effective by themselves, but their value can be increased if the services provided are useful.

This is the so-called added effect of platform goods. There is a further, relative increase in the value of platform goods if the group accepting and using the platform goods grows larger. This is called the share dependency effect of platform goods. In the world of platform goods, the accumulative action of these two effects changes the choice structure of those who demand the platform goods according to their shares. This quickly constructs a monopolistic or oligopolistic structure in platform goods. The monopoly caused by this active change in demand-side effects is called a demand-oriented monopoly [Deguchi, 2000b].

9.2 Platform Externality

The concept of platform goods plays an important role in digital society. As we have pointed out we analyze interactions among three groups, namely providers of goods called platform goods, providers of service good that can be provided based on this platform, and their consumers.

We introduce the concept of platform externality which is an extension of the concept of network externality. Network externality is a classical notion for competition among networks, where the utility of goods or services on a network depend on the market share of the network. The notion is used for analyzing the competition between standards such as VHS vs. Betamax [Katz, 1986; Cornes, 1996]. We extend the notion of network externality to the vertical reliance between service goods and its platform. We assume two types of platforms such as α and β in Fig. 9.3.

In this chapter we focus on companies who provide service goods on the platforms and on consumers who buy both platform and service goods. We also analyze the competition among companies who provide platforms. We show an example of platforms and service providers on the platforms, which are shown in Fig. 9.4. In the example α means Macintosh operating system and β means Windows operating system. In another example α may mean Sega Dreamcast game machine and β may mean Sony PlayStation game machine.

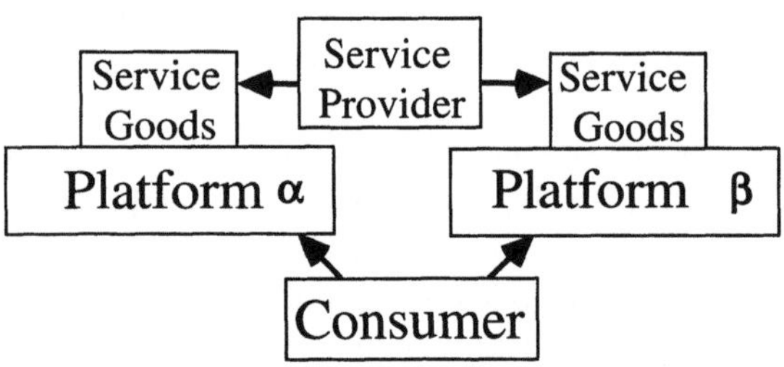

FIG. 9.3 Platforms and Service Goods on the Platforms

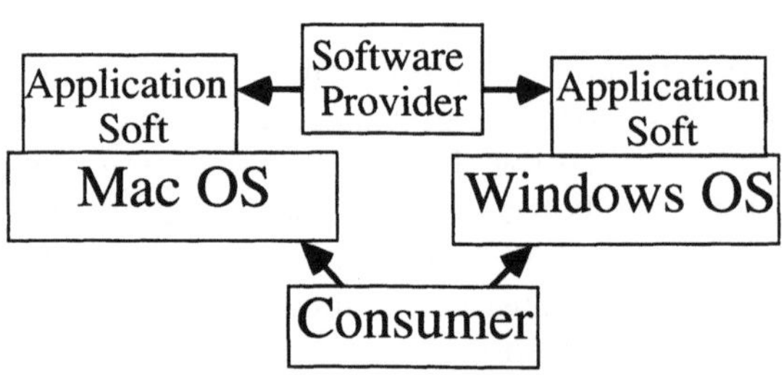

FIG. 9.4 An Example of Platforms and Services on the Platforms

This section conceptually studies cases that enable an analysis of the two dynamics: an alternative chosen by consumers of platform goods and an alternative chosen by providers of service goods. The analysis is made based on the social learning dynamics of decision-making, which is given in the next section.

In the model consumers and service providers change their attitude slowly under the influence of the average payoff of alternatives. The average payoff might be obtained by the random matching game in the context of evolutionary game theory. But in this case we assume that the average payoff is obtained by macro social and economic information from such macro social sectors as the the mass media. We especially pay attention to these social learning processes of autonomous agents under macro information.

9.3 Model of Platform Externality and Lock In

In the model, a player changes his attitude slowly under the influence of average payoff of alternatives. In other words we derive the dynamics not from direct interaction among agents and its payoff, but from single decision making depending on the average payoff of alternatives.

There exist two groups of players, related to two platforms. One is a group of companies that provide service on the platforms and are called service providers. The other is a group of consumers that buy both a platform and service goods on the platform.

The service providers have the three alternatives. "b_1" , "b_2" and "b_3" which means that the companies provide service goods on the platform α, β and both respectively. The consumers have the three alternatives "a_0", a_1 and a_2, which denotes the initial condition that they have nothing and the selection of platform α and β respectively. These are shown in Fig. 9.5.

Model construction occurs from the average payoff of alternatives. Table 9.2 shows the average payoff of consumers for each alternative. The expressions given in Table 9.2 are defined as:

$P[b_1]$ or $P[b_2]$ are the population ratio of service providers that provide service goods only on the platform α or β respectively.

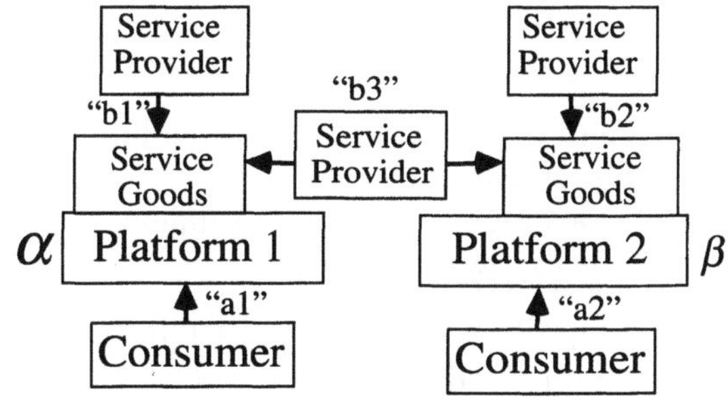

FIG. 9.5 Alternatives for Platform Selection and Service Selection

TABLE 9.2 Average Payoff for Alternatives of Consumers

Average Payoff	From Service Goods	From Platform Goods
$E[a_0]$	0	0
$E[a_1]$	$K_{10} + K_{11}P[b_1] + K_{11}P[b_3]$	$h_{10} + h_{11}Q[a_1] - d_1$
$E[a_2]$	$K_{20} + K_{21}P[b_2] + K_{21}P[b_3]$	$h_{20} + h_{21}Q[a_2] - d_2$

TABLE 9.3 Average Payoff for Alternatives of Service Providers

Average Payoff	Benefit from Selling Service Goods	Cost from Providing Service for Platform
$E[b_1]$	$r_{10} + r_{11}Q[a_1]$	$-C_1$
$E[b_2]$	$r_{20} + r_{21}Q[a_2]$	$-C_2$
$E[b_3]$	$r_{30} + r_{11}Q[a_1] + r_{21}Q[a_2]$	$-C_3$

$P[b_3]$ is the population ratio of service providers that provide service goods on the both platforms.

$Q[a_1]$ or $Q[a_2]$ are the population ratio of consumers that select the platform α and service goods on α or the platform β and service goods on β respectively.

d_1 is the cost to use platform α per period.
d_2 is the cost to use platform β per period.

h_{10} is the technological utility of platform α.
h_{20} is the technological utility of platform β.

$E[a_1]$ is the average payoff of consumers that select platform α per period.

$$E[a_1] = K_{10} + K_{11}P[b_1] + K_{11}P[b_3] + h_{10} + h_{11}Q[a_1] - d_1$$
$$E[a_2] = K_{20} + K_{21}P[b_2] + K_{21}P[b_3] + h_{20} + h_{21}Q[a_2] - d_2$$

Table 9.3 shows the average payoff of service-providing companies for each alternative. From Table 9.3,

$E[b_1]$, $E[b_2]$ and $E[b_3]$ are the average payoff of service provider that select platform α, β and both per period respectively.

$$E[b_1] = r_{10} + r_{11}Q[a_1] - C_1$$
$$E[b_2] = r_{20} + r_{21}Q[a_2] - C_2$$
$$E[b_3] = r_{30} + r_{11}Q[a_1] + r_{21}Q[a_2] - C_3$$

C_1, C_2 and C_3 are the cost for providing service goods on the platform α, β and both respectively.

Total average payoff of consumers, $E[a]$, and of service providers, $E[b]$, are shown as follows.

$$E[a] = Q[a_0]E[a_0] + Q[a_1]E[a_1] + Q[a_2]E[a_2]$$
$$E[b] = P[b_1]E[b_1] + P[b_2]E[b_2] + P[b_3]E[b_3]$$

Where $Q[a_i]$ and $P[b_j]$ denote the population ratio for each alternative in consumers and service-providing companies respectively. Then we deduce the following social learning dynamics [Deguchi, 1998; 2000a,b; 2003b].

For Consumers:

$$dQ[a_1]/dt = \{Q[a_1](E[a_1] - E[a])\}/E(a)$$

$$dQ[a_2]/dt = \{Q[a_2](E[a_2] - E[a])\}/E(a)$$

$$Q[a_0] = 1 - Q[a_1] - Q[a_2]$$

For Service Providers:

$$dP[b_1]/dt = \{P[b_1](E[b_1] - E[b])\}/E(b)$$

$$dP[b_2]/dt = \{P[b_2](E[b_2] - E[b])\}/E(b)$$

$$P[b_3] = 1 - P[b_1] - P[b_2]$$

This model includes two types of externalities. One comes from the sharing of a platform among consumers and the other from the sharing of a platform among service providers. We add small fluctuations in the simulation. C_3 is the cost for providing service on the both platforms. Thus $Max(C_1,C_2) \le C_3 \le C_1 + C_2$ holds. We assume $C_1 = C_2 = C$ and $C \le C_3 \le 2C$ for simplification. h_{10} and h_{20} show the technological utility of the two platforms respectively. d_1 and d_2 denote respective cost of the platform α and β, for consumers per period.

The simulations are shown as follows. We assume initial conditions about the market share of platforms by consumers as $\alpha = 11\%$ and $\beta = 10\%$ in Figs. 9.6 and in 9.7. We also assumed $C_3 = 1.7 \times C_1$, $C_1 = C_2 = 0.1$ in Table 9.3 and $h_{20} = h_{10} = 0.1$ in Table 9.2.

Figures 9.6 and 9.7 show the effect of the fluctuation on initial share. As a result $Q[a_2]$ and $P[b_2]$ become unity under the small difference of the initial share condition. $Q[a_1]$ and $P[b_1]$ may become unity under the small perturbation.

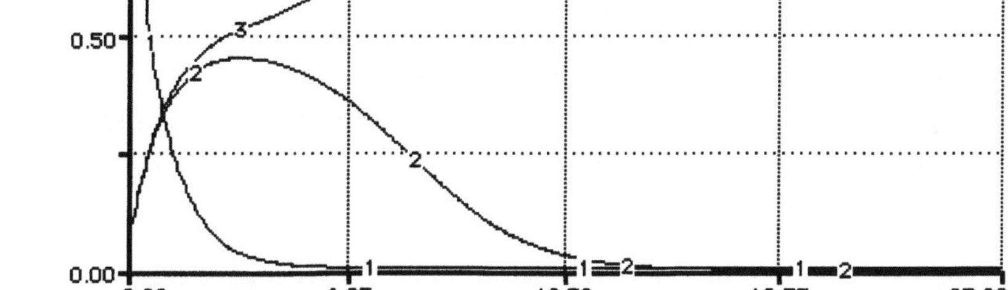

FIG. 9.6 Platform Selection by Consumers

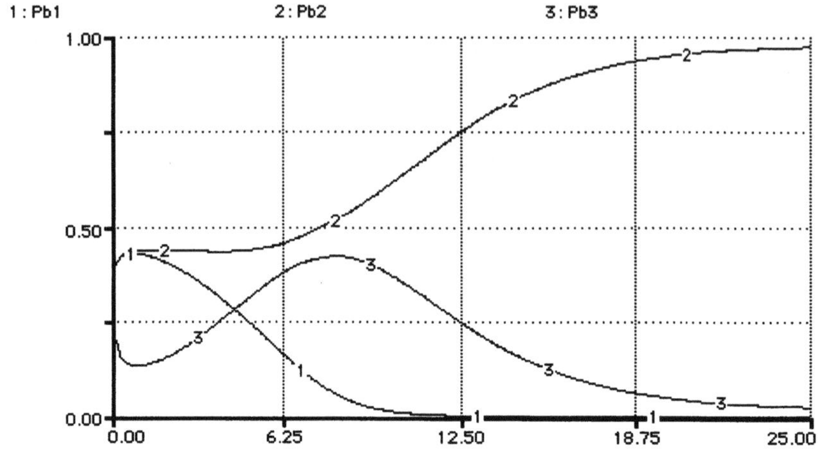

FIG. 9.7 Platform Selection by Service Providers

Strong lock-in occurs with the monopoly coming from the dynamic change of demand depending on the platform share of consumers and service providers and on their mutual interaction. This type of monopoly does not arise from supply side effects of platform-providing industry such as economy of scale or scope. Rather, it comes from the dynamic change of utilities of the consumers and service suppliers on the platform and their mutual learning dynamics. Thus we call this demand-oriented monopoly.

The results are shown in Fig. 9.8 and Fig. 9.9.

No Crossover Cost Case:
If there is no crossover cost then lock-in still occurs in the process of consumer's selection but the tendency is weaker than the previous case with crossover cost. Furthermore there is an incentive for service providers to provide service goods to the both platforms in this case. Thus there is a time for looser platform provider to apply a new counter strategy. It is important to reduce crossover cost by a certain industrial policy to maintain fair competition depending not on the previous strategy but on the current effort of the firms.

The results are shown in Fig. 9.8 and Fig. 9.9.

No Crossover Cost and No Share Effect Case:
If we can ignore not only the crossover cost but also influence of share as bandwagon effect then the competition becomes more effective. The results are shown in Fig. 9.10 and Fig. 9.11.

Release of Lock-in by Change of Technological Utility and Price:
Looser platform provider can come from behind and win even in the lock-in situation if the technological utility of the looser platform becomes high enough

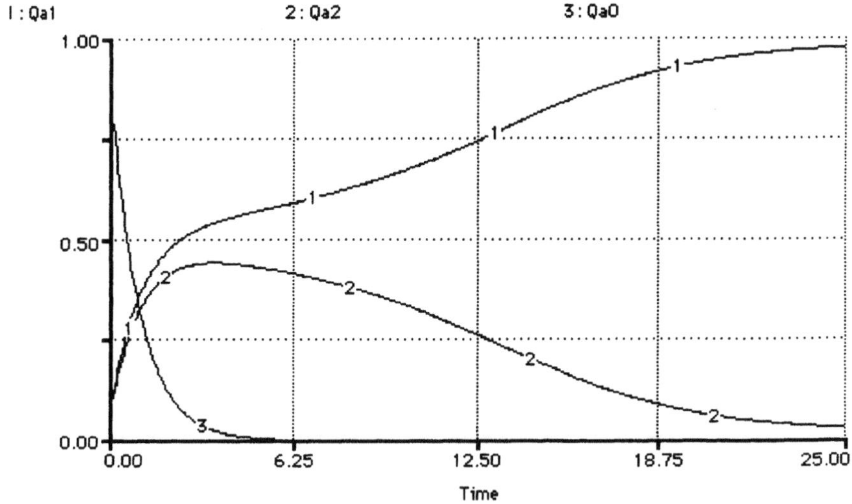

FIG. 9.8 Platform Selection by Consumers with No Crossover Cost

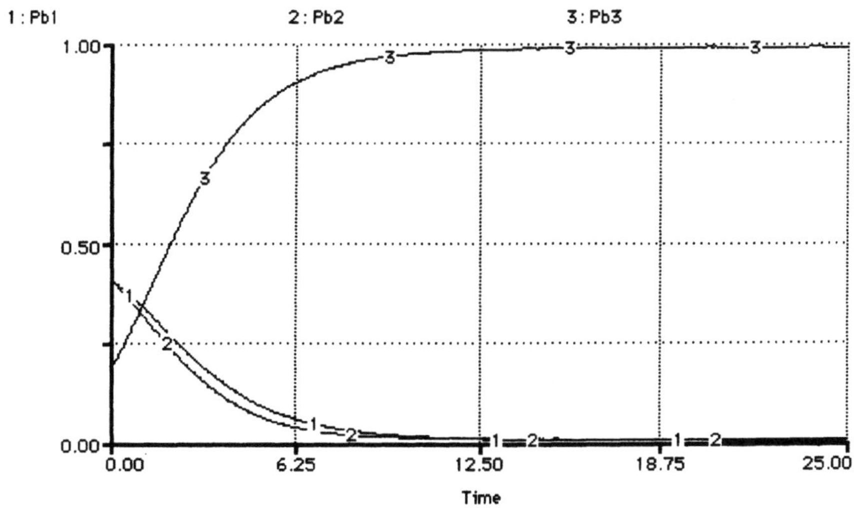

FIG. 9.9 Platform Selection by Service Providers with No Crossover Cost

and cost to use the platform becomes low enough. Figs. 9.12 and 9.13 show the case that the lock-in is released.

However the competition is not fair for the looser platform provider, once the lock-in occurs, because extra effort is required for the looser compared with no lock-in market. Thus the diffusion of new technology might be prevented under this unfair competition. The unfair condition is unavoidable from the platform

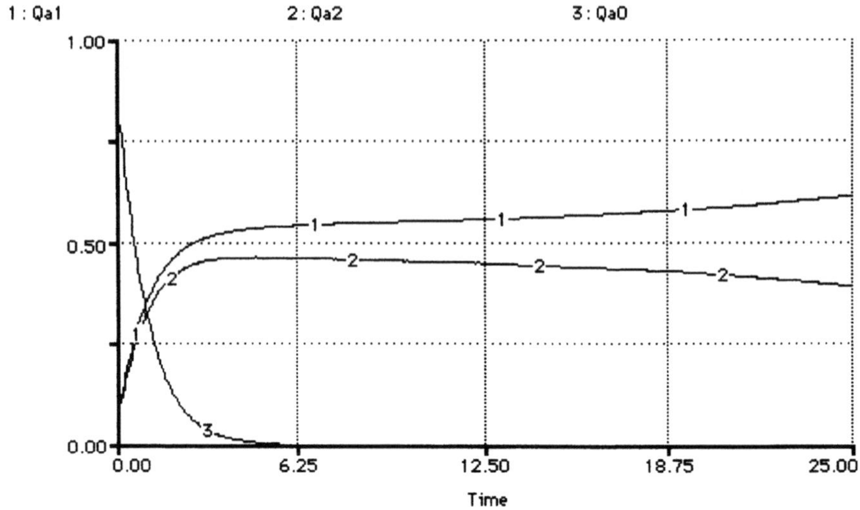

FIG. 9.10 Platform Selection by Consumers with No Crossover Cost and No Share Effect

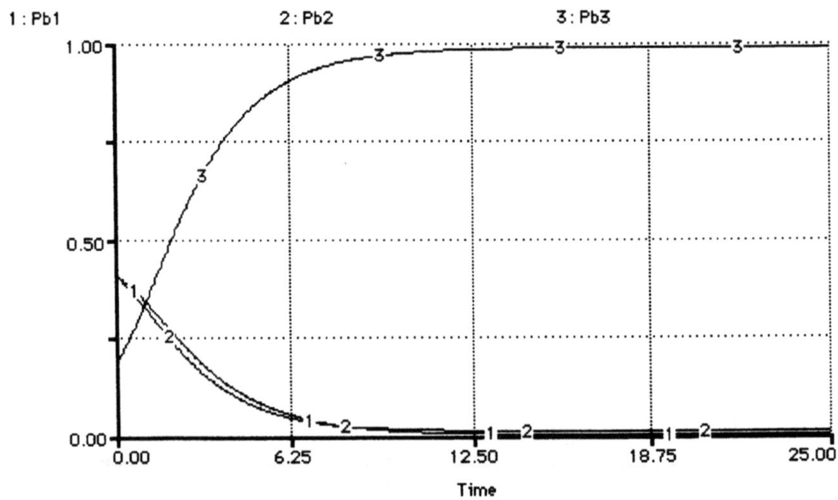

FIG. 9.11 Platform Selection by Service Providers with No Crossover Cost and No Share Effect

structure. Then there happen to be unfair gap between competitors. The problem is discussed in the following section by using bifurcation analysis.

Network Reputation:
Network reputation might change lock-in situation. We have already shown the effect of support by knowledge in Chapter 3. As the same way we introduce the

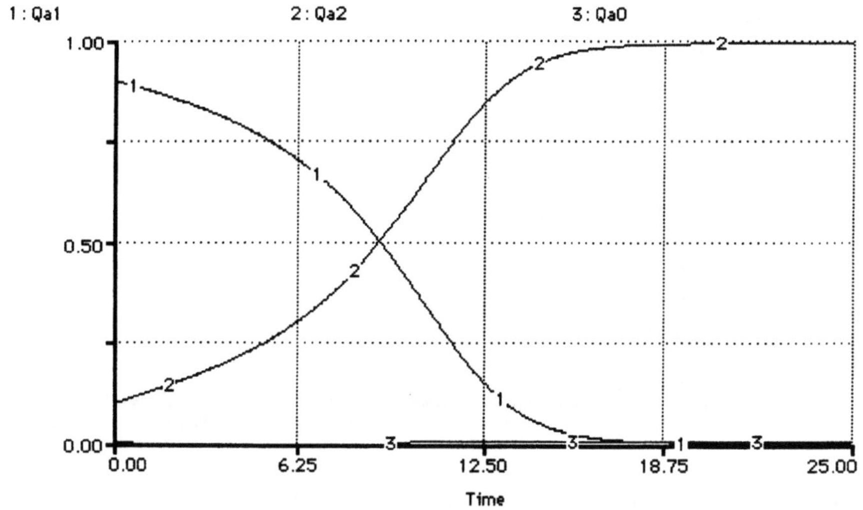

FIG. 9.12 Release of Lock-in at Platform Selection by Consumers

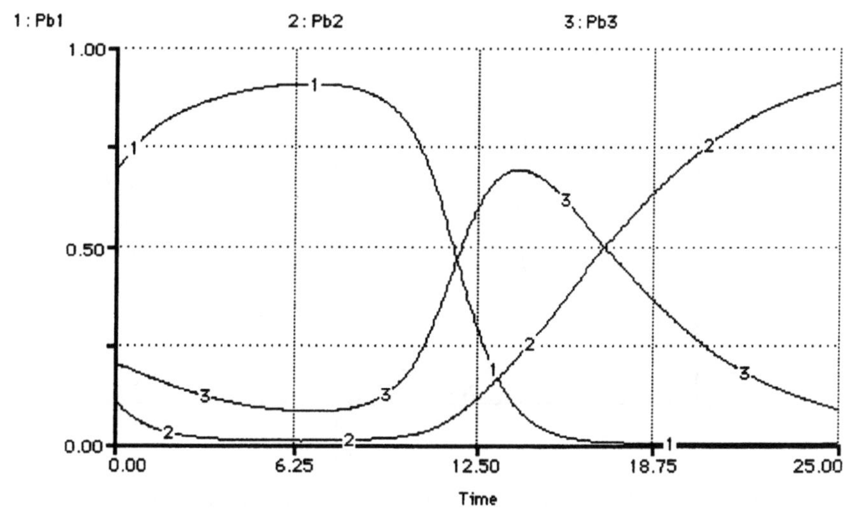

FIG. 9.13 Release of Lock-in at Platform Selection by Service Providers

effect of network reputation as support by knowledge. We give simple assumptions insisting if share of one product increases then the reputation decreases as follows.

$$EAKnow = NetEvalA$$

$$EBKnow = NetEvalB$$

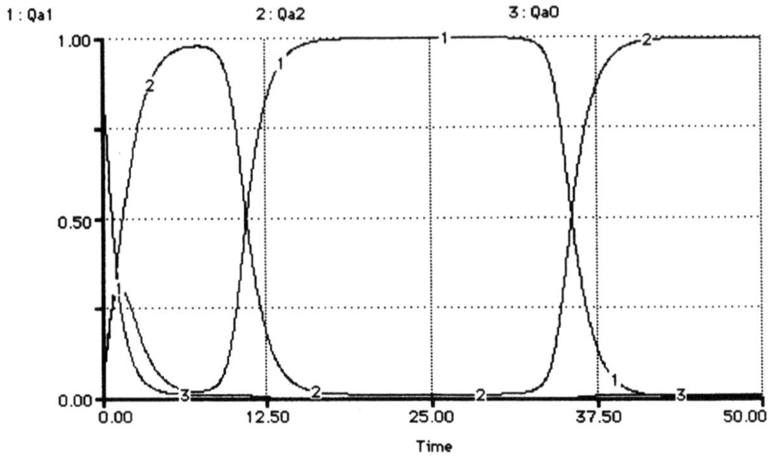

FIG. 9.14 Platform Selection by Consumers under Network Reputation

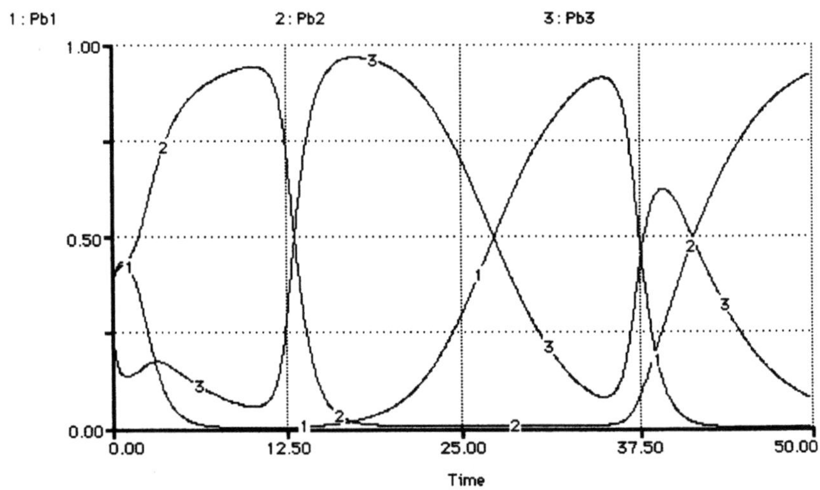

FIG. 9.15 Platform Selection by Service Providers under Network Reputation

$$\text{EASupport} = \text{PaKnowledge}$$

$$\text{EBsupport} = (1\text{-PaKnowledge})$$

$$\text{NetEvalA} = P[b_2] + Q[a_2] \times 0.2$$

$$\text{NetEvalB} = P[b_1] + Q[a_1] \times 0.2$$

As a result the share oscillates as shown in Figs. 9.14, 9.15 and 9.16.

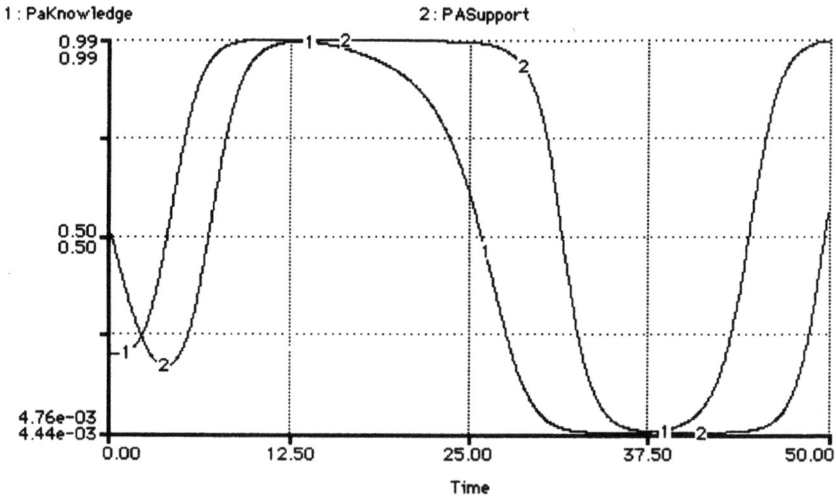

FIG. 9.16 Change of Knowledge and Supporting Attitude

9.4 Bifurcation Analysis of Social Learning Dynamics

Next we analyze the bifurcation process of the social learning dynamics.

$$dQ[a_1](t)/dt = Q[a_1]\{E[a_1] - E[a]\}/E(a) = Q[a_1](1 - Q[a_1])\{E[a_1] - E[a_2]\}/E(a)$$

The stability of this dynamics depends on the sign of the term $\{E[a_1] - E[a_2]\}$.

$$\begin{aligned} E[a_1] - E[a_2] = {} & K_{10} + K_{11}P[b_1] + K_{11}P[b_3] + h_{10} + h_{11}Q[a_1] - d_1 \\ & - K_{20} - K_{21}P[b_2] - K_{21}P[b_3] - h_{20} - h_{21}Q[a_2] + d_2 \end{aligned}$$

Let $h_{21} = h_{11} = h_1$, $k_{11} = k_{21} = k_1$, $K_{10} = K_{20}$.

If there is no competition of cost and technology between platform-providing companies then it can be assumed that $d_1 = d_2 = d$, $h_{20} = h_{10} = h_0$.

$$\begin{aligned} \text{Then } E[a_1] - E[a_2] &= h_1\{Q[a_1] - Q[a_2]\} + k_1\{P[b_1] - P[b_2]\} \\ &= h_1\{2Q[a_1] - 1\} + k_1\{P[b_1] - P[b_2]\}. \end{aligned}$$

Let $P[b_1] = P[b_2] = 0$. This means there is no additional cost for providing service goods to both platforms.

Then $E[a_1] - E[a_2] = h_1\{2Q[a_1] - 1\}$ holds. $Q[a_1] = 1/2$ is an unstable equilibrium point. $Q[a_1] = 0$ and 1 are a stable solution. Thus the share of platforms is locked in the long run depending on the initial share. Nevertheless the competition is said to be dynamic and fair if price (d_1, d_2) and technology (h_{10}, h_{20}) have an effect on the competition.

Now we consider the competition of price and technology between platform-providing companies for platform selection. If we assume that there is no crossover cost for providing service goods to both platforms, i.e. $C_3 = C_1 = C_2 = C$. Then $P[b_1] = P[b_2] = 0$ holds.

[Proposition 9.1]
$E[a_1] - E[a_2] = h_1 \times \{2Q[a_1] - 1\} + k_1 \times \{P[b_1] - P[b_2]\} + d_2 - d_1 + h_{10} - h_{20}$.
Let $\Delta = d_2 - d_1 + h_{10} - h_{20}$ and $C_3 = C_2 = C_1 = C$. Then

(1) If $(h_1 - \Delta)/2h_1 \leq 0 \Leftrightarrow h_1 \leq \Delta$ then $Q[a_1] = 1$ is stable.
(2) If $1 \leq (h_1 - \Delta)/2h_1 \Leftrightarrow \Delta \leq -h_1$ then $Q[a_1] = 0$ is stable.
(3) If $0 < (h_1 - \Delta)/2h_1 < 1 \Leftrightarrow -h_1 < \Delta < h_1$ then $Q[a_1] = (h_1 - \Delta)/2h_1$ is unstable.

Proof:
$E[a_1] - E[a_2] = h_1 \times \{2Q[a_1] - 1\} + k_1 \times \{P[b_1] - P[b_2]\} + \Delta$.
$= 2h_1 \times Q[a_1] - h_1 + \Delta$ holds. Thus
$Q[a_1] = 1$ is stable $\Leftrightarrow 0 \leq E[a_1] - E[a_2] \Leftrightarrow (h_1 - \Delta)/2h_1 \leq 0 \Leftrightarrow h_1 \leq \Delta$
$Q[a_1] = 0$ is stable $\Leftrightarrow E[a_1] - E[a_2] \leq 0 \Leftrightarrow 1 \leq (h_1 - \Delta)/2h_1 \Leftrightarrow \Delta \leq -h_1$
$Q[a_1] = (h_1 - \Delta)/2h_1$ is unstable $\Leftrightarrow 0 \leq (h_1 - \Delta)/2h_1 \leq 1 \Leftrightarrow -h_1 < \Delta < h_1$

Q.E.D.

Thus $Q[a_1] = (h_1 - a)/2h_1$ is an equilibrium point.

Let $h_1 = 0$ then the stable point will change depending on the sign of Δ ($\Delta = d_2 - d_1 + h_{10} - h_{20}$), which is determined by pure competition over cost (price) and technology. $d_2 - d_1$ represents price competition and $h_{10} - h_{20}$ represents technology competition.

If the competition over price and technology change the sign of Δ then the stable equilibrium point will change, bifurcated by the competition. This means that competition is fair and heal.

In some cases there is a strong barrier against fair competition. If h_1 is not zero then there is a barrier to fair competition that has $2h_1$ in depth. The barrier is caused by platform externality. In this case the coexistence of platforms is difficult to achieve because of its externality. If crossover costs exist, then the barrier becomes higher.

9.5 Platform Goods and Technological Innovation

9.5.1 Technological Innovation and the Product Lifecycle

Christensen divides corporate technological development into: (1) innovation by sustaining technologies, which tends to generate more elaborate specifications than necessary by advancing the existing technology to cater to customers' needs in the existing market; and (2) innovation by destructive technologies that is generated targeting a certain market but ends up controlling a different market [Christensen, 1997]. Advanced technology views this destructive technological innovation initially as merely a technology available for low-end products, but eventually it makes inroads into the high-end of the market.

Apart from the discussion of this technological innovation, there is a problem unique to a market that recognizes platform goods and the service goods based on those platform goods as a combination. To introduce goods and services based on new technology to this kind of market, it is necessary to consider the following aspects:

(1) the relationship between redemption of the previous capital investment and new capital investment
(2) whether the market has become saturated with old products, or whether there is a cannibalization of old products caused by the market shift
(3) the issue of the change in competitive structure by dissembling the lock-in structure formed on the existing platform.

We will use these cases to find out specifically what arises in the market from the technological innovation that accompanies this platform shift. What we focus on is the strategy to deliberately delay technological innovation. The technologies we refer to and focus on here by technological innovation are not the technologies accumulated by a company inside the organization as a result of R&D investment; rather they are the technologies that introduced the above technologies to the market.

Making an investment in R&D and introducing the result to the market are two different things. Basically, if a product produced using a new technology is sufficiently competitive, then introducing this product to the market will write off the capital investment in the old products and that would be an appropriate choice, especially if the market for the old products is becoming saturated. In the platform-type industrial structure, however, a new element is added as a factor in any judgment as to whether to introduce the result of R&D to the market or not. The consideration of the capacity of manufacturing technology to improve productivity is excluded here.

Home electronic products, such as refrigerator, are supplied in a single unit functionally independent and its technological innovation is directly linked to competitiveness. If the considerations are given to the above issues (1) and (2), the new product is to be introduced on the market. However, in the aspects of the actual innovations, especially in the case of high-technology products, various dilemmas are generated and may cause interesting dynamics when something called destructive innovation invades the existing market, as Christensen pointed out. Christensen explains that a destructive technology is evaluated in the new value network and market and starts invading the existing market at a blistering pace once it reaches a certain stage.

What we found in the market where platform goods and service goods based on the platform goods are combined is the suspicion that the process and effects of destructive innovation are eased in the field of final consumer goods and the speed of technological innovation decreases.

If this is true, there is a possibility that its business strategy evaluation will not match the macro politic evaluation. A market that is locked-in to a specific platform has a symbiotic relationship between service goods and platform goods being formed and makes it difficult for other platforms to enter. We will verify the impact of this on technological innovation by using specific examples.

9.5.2 Examples of Platform Goods and Technological Innovation

(1) Electric Power Network

The electric power industry is a platform industry whose platform is the power transmission network and whose service goods based on this platform are power generation services. The industry formerly had a structure in which companies providing power transmission platforms and companies providing power production services were vertically integrated.

The innovation of cogeneration technology in this market had been carried out with extreme deliberation. Despite this, since the power production service was deregulated, allowing new power production services to enter the market as service goods, cogeneration technology has achieved significant development but having been centered on large-scale power generation services, the electric power industry is still not keen on purchasing power from wind-power or cogeneration.

To see a difference in the patterns of technological innovation or in corporate competition between a platform industry and a non-platform industry, we will compare the electric power industry and the steel industry.

The steel industry does not have a platform type structure, apart from sales channels. Because of this, when electric furnace appeared as a destructive technological innovation for downsizing, there was nothing to prevent it from technological innovation and from entering other markets. Electric furnace has become a technology comparable to blast furnaces today, by retaining its own niche and then by quickly achieving technological innovation. This is a typical example of destructive technological innovation.

Innovation like this did not occur in electric power facilities in the electric power industry. This is because electric power is a platform industry that needs the power transmission that serves as its platform.

(2) ADSL Service

Internet connection services by provider can be divided into services to provide the Internet itself and a service to provide access lines. The combination of these two services becomes a platform to provide and use the services on the Internet. The Internet connection service using ADSL employs the existing telephone line connection service as its platform and, at the same time, it is a platform in itself for the Internet use.

Tokyo Metallic Communications Co., Ltd. opened up the ADSL market in Japan, but failed to obtain enough connection lines and was eventually taken over. What underlies this incident is the issue of the vertical integration of platform goods and service goods. NTT, which provides telephone line connection services, has ample technologies in ADSL, but the company has no positive incentive to set up an ADSL market at that time because ADSL would only compete with the ISDN Internet connection and NTT has not finished writing off its investment in the ISDN network.

SOFTBANK Corp. has succeeded in cultivating the ADSL market with its aggressive low prices. NTT, which has a hold of the platform, namely the telephone line connection service, coincidentally entered the new market with perfect timing and with minimum risk. It is now well on its way to grabbing a large share of the market.

While NTT was virtually bottling up the ADSL technology inside the company or delaying the entry of other companies to the market as much as possible, once it saw that the ADSL market had developed, it succeeded in acquiring the market by making a sudden investment in it.

(3) Game Machine Market and Competition among Business Models of the Market

Game hardware is typical of platform goods. Nintendo Co., Ltd. had a strong impact in the early days of this market and its strategies were very distinctive. Against the backdrop of the success of the Family Computer, Nintendo orchestrated game software companies have gone from strength to strength and at the same time, it succeeded in establishing a strong distribution platform.

As a result of this, Family Computer was able to accumulate software providers on its platform and establish a competitive edge as a platform good. A strategy of enclosing software goods within its platform to improve the attractiveness of the platform is common to CDs, videos and DVDs and is not so unusual. A strategy of accumulating software goods on the company's platform to obtain competitiveness is attracting greater attention, as platform goods become more widespread.

Since killer software splits the competition among platforms in the game industry, game manufacturers themselves have been vigorously developing killer software, just like Nintendo developed Pokemon and Super Mario Brothers. On the other hand, the companies providing software closely watch the competition among platforms to decide which platform to provide their software to and then make the switch if necessary.

Once Nintendo succeeded in transferring to a new generation, the 16-bit Super Family Computer, from the Family Computer, it adopted a strategy of intentionally delaying the transfer to a next generation game expected to have large-capacity storage media, such as CD-ROM. This does not, however, mean that Nintendo reduced its investment in R&D. It developed and introduced to the market a game machine using 3D glasses during this period, although it was not successful. It also developed a new product like Game Boy using a new technology and successfully introduced it to the market. This means it is possible to make vigorous R&D investments while bottling up a technology in the market.

There is a risk that an advantage of externality established by one's own platform may be lost once the platform is superseded by the next generation platform.

Where there is an unfair gap as a barrier that stops other platforms from entering the market, a company would not have an incentive to vigorously promote the transfer to the next generation platform. It is more effective to bottle up a technology. In the game market, Sony Corporation has succeeded in converting the platform by changing the conventional business model as a whole with PlayStation, which enabled it to cross over the boundary of the platform externality. This success is owed to new incentives given to software manufacturers to make the switch to its platform, e.g. opening up a new sales channel with CD-ROM, reducing stock risk and providing a reasonable development environment.

(4) Platform of Distribution Industry

Some of the old distribution mechanisms had platforms where dealerships exclusively dealt in products of a specific manufacturer and products and services were not provided by crossing over the borders of these dealerships. Japanese automobile dealerships were once severely criticized by the United States for being unfair, as its sales channels were vertically integrated by manufacturers.

It is costly to maintain a distribution platform involving dealerships. Thus dealerships were vertically integrated by automobile or home electric appliances manufacturers and became a platform for supplying only that manufacturer's products. For this reason, once the market experiences a cataclysm, these dealerships become the weak point of their manufacturers. While Matsushita Electric Industrial Co., Ltd. is struggling to reform its business model, which is saddled with a long list of dealerships, a company like Sony, thought to have a weak distribution network, achieved innovation in its business model and is now developing its distribution channel using the Internet with the Sony-style brand.

If an exclusive platform where distribution is vertically integrated is an unfair business criticized by the United States, it is necessary for restraining various platform goods that have similar structures from vertically integrating with their service goods.

(5) Case Study of the Photocopier Market

Xerox Corporation had a substantial share of the photocopier market. This is because the photocopier market is a service providing industry that has a maintenance network as its platform and latecomers could not penetrate this vertically integrated service network.

But a maintenance-free photocopier developed by a Japanese company made inroads into the market from the lower end of the market as a destructive technology. Xerox's share dropped significantly in a market where the maintenance service-oriented business model was now competing with the maintenance-free business model that used a drum-toner hybrid. This is an example of business model innovation crossing the entry barrier formed by platform externality.

(6) Microsoft OS and Network Strategy

An operating system is a type of service goods using PC hardware as its platform goods. At the same time it is a platform good for application software. The most appropriate strategy for a platform in the center of this platform hierarchy is:

(1) for lower platforms, promoting competition among the lower platforms by providing services uniformly to as many platforms as possible
(2) for higher platforms, accumulating as many service goods on the company's own platform as possible

Microsoft Corporation supplied its OS to the machines of numerous manufacturers, although all within the common framework of the IBM PC specification. As a result, it has acquired its own hardware platform that evolved rapidly without it having to make any development investment.

Apple Computer, Inc. lagged in technological innovation since it vertically integrated its OS with its hardware and it had to evolve its hardware by making its own investment. An example of this can be seen in the history of the downsizing of its computers, from the laptop, to the notebook, to the sub-notebook. Apple, which had a head start on laptops in the early days, left the others in its wake in the technological development stakes. As a result, Microsoft could add the effect of compact hardware to the effects of its own OS without making any development investment.

Microsoft has been intentionally driving forward a number of different strategies by using the platform externality in this OS market. It attempted to make MSN the main platform for email services, etc. to take on the Internet by integrating its connection service to Microsoft Network Service (MSN) into the OS in the embryo stage of the Internet. However, a destructive technological innovation, the Internet, soon swept aside the BBS market.

On the other hand, Microsoft pursued aggressively a strategy of expelling any technologies, such as JAVA of Sun Microsystems, Inc. and RMI, which is one of the key technologies of distributed object oriented technologies that could put the concept of its OS in danger, from the market. For this purpose, it localized Java on its own Windows platform and tried to penetrate the market with a better version of RMI. In a market where the OS is locked in, its local version Java would become the de facto standard.

This issue has been argued in the Microsoft antitrust battle. The company had no choice but to abandon its strategy. However, when this lawsuit was settled, it made a decision to remove the JAVA virtual machine from Windows XP.

9.5.3 The Technological Innovation Strategy of Platform Providing Companies

(1) Competition of Service Goods

When service goods that are vertically integrated on certain platform goods need to compete with the service goods of another company, the company providing the platform goods has a strong incentive to limit the provision of this platform in various ways, to make things difficult for another company's service goods.

Unless this company can provide the equivalent service or if providing this kind of service will compete with a service already provided by the company, it is possible that it will adopt a strategy of limiting the use of its platform for various reasons. This was seen in wind-power generation, cogeneration and the communications market. As a result of this, these competing technologies face the possibility of being locked out of the market because they were introduced too late.

On the other hand, if the entry of competing service goods to the market is unavoidable, it is possible to adopt a strategy called the "back eat & catch up strategy". When powerful service goods, which the company has been overlooking or thinking of driving out, arise in the market, it is easier for the company to catch up all at once by using the platform's externality and looking for the right time to enter the market.

The ways that Microsoft entered the streaming technology and browsing technology markets and NTT entered the ADSL market can be regarded as good examples of using this kind of technology approach.

(2) Competition among Platform Goods

For the competition among the platform goods themselves, the basic strategies would be as follows: (1) Aiming to lock-in market share by accumulating software goods on the platform; (2) Delaying as much as possible the transfer to the new generation platform with successful externality to maximize profits. These strategies will generally reduce the speed of technological innovation of the platform goods. (3) A technology that could threaten the platform generates an incentive to drive it out by using the externality of the platform of the company as much as possible. This was seen as Microsoft's strategy of localizing JAVA.

If a transfer to a next generation platform or the introduction of a new technology that could threaten an existing platform advantage is unavoidable, (1) trying to buy as much time as possible and (2) approaching the new market by using, from the opposite direction, the advantages of the service goods vertically integrated in the company's platform.

The responses of Microsoft to the Internet have evolved in the following way:

(1) In the beginning, it attempted to kill off the Internet, if possible, and to make its own MSN the network platform by using the externality of its OS.

(2) It then acknowledged the importance of the Internet and vertically integrated the important service applications of the Internet in its own OS platform. For this purpose, it adopted the back eat & catch up strategy.
(3) It prevented the penetration of destructive innovation, e.g. Java and distributed object-oriented technology. It used its own platform externality to make its technology more advantageous. Open source also became a target to be eliminated.
(4) To shift its pivoting foot from OS to a more destructive innovation, the distribution object-oriented technology, it introduced a technology developed during the time being bought.

It also aimed at gaining a platform for business system construction on the network for the next generation network by getting maximum leverage from the advantages of the current platform.

In this way, as seen in the above steps, Microsoft, as a platform providing company, has been using strategies to seal and control the technologies while making a huge investment in R&D.

Destructive innovation is something that is difficult to predict and hard to respond to quickly. There is a strong incentive to restrict another company's destructive innovation itself. Regarding this incentive, an industry that has platform goods has several effective strategies, as described above.

A company that acquired a share in platform goods can obtain an advantage in technological innovation or investments in service goods on top of its platform, if it pursues this kind of strategy. Many companies providing platforms do in fact appear to be carrying out the above strategies intentionally, or at least extremely ingeniously.

9.5.4 Institutional Design of a Platform-type Industry Structure

There seems to be a consistent tendency in platform goods to slow the transfer to the next generation technology. As long as an unfair gap with a dependency on the share of platform exists in competition, there is a self-evident entry barrier. The conventional antimonopoly policy is petty in terms of inhibiting technological innovation. This is because the theory on which it is based focused on price control by a market oligopoly or monopoly.

Our analysis indicates that the policy to divide a company regionally in the power industry or the telecommunications industry may not necessarily be effective for technological innovation if the vertical integration of platform goods and its service goods are left ungoverned.

In the digital economy under development, the platform functions of the existing distribution channel have been destroyed and this is enabling new business models to develop. On the other hand, a new link between large-scale platform type goods and services is being formed, centering on information goods.

In a market like this, existing institutional regulations are not good enough to attain their objects. Institutions need to have a continuous process of feedback where they are constantly verified and redesigned by referring to higher objectives in the changing system. Our discussion indicates that it is essential to discuss new institutionalization like this in order to design the digital economy.

9.6 Cultural Platforms

9.6.1 Introduction

The social space of the real world has been constructed gradually throughout history and is in need of some kind of management to respond to globalization today. This real social space consists of elements such as various objects that make up everyday life, activities associated with these objects, and interpretations and evaluations of these objects. The various objects in this everyday life world are externally referred to, evaluated and selected by all kinds of bodies and there is an encouragement to learn, evolve and improve them.

Places for mutual reference and evaluation in modern times are systems consisting of markets, corporate organizations, governments (administration), regions, the mass media, and other components. Recently, however, a kind of medium system, like the Internet, has been formed and is playing an active role. Moreover, the information network is about to play a critical role in forming this medium group and in mutual support.

Today's social system is not something that can be interpreted only within the market or within the nation-state or within international relations. For management of the cultural and national world, it is necessary to question what kind of place needs to be constructed as a place for the reproduction of "reality" in this world from the standpoint of design theory of everyday life. This section examines the requirements for a social system to enable multifactorial worlds to coexist, through an analysis of the groups responsible for social realities.

9.6.2 Language and Culture as Platform Goods

The Modern Era began when people stopped being the lambs of God and took the power to evaluate their own actions around individual axes. However, the actual system was constructed in the market, which serves as a place for objective evaluation, and government, which serves as a supplement to the market. However, this system is on its way to being diversified and is now changing rapidly. Moreover, in the case of cultural objects and information communication goods, their evaluation and selections are often subject to dynamic changes caused by demand-side factors, such as the number of people using these goods. This often brings a situation where the weak become victims of the strong.

This mechanism is the same type as applies to our analysis of platform goods. The concept of platform can also be formed regarding cultural objects beyond goods and services as products. For example, language is a typical example of platform goods since diverse knowledge is provided with it and it is a place to be used. Currency can also be regarded as a form of platform goods since it enables the exchange of services by using it. Language and currency have a share dependency effect with respect to the convenience of their use. In the progress of informatization, our culture may lose its multifactor nature and be locked in to a single world vision.

A language, in particular, serves as a very important cultural platform. Platform goods form a hierarchy, while a unit of a cultural set based on a language serves as a platform that provides a set of world visions to live by on this platform. People constitute a system that can function when platforms like language and culture are installed in their body (wet wear platform). To ease the monopoly of these linguistic cultural platform goods, it is extremely important to evenly distribute the service goods available on these platforms and to find unique value in each platform.

Bank notes are platform goods, but a diversity of currencies are maintained by freely carrying out exchanges between different notes. English is now about to gain superiority as a linguistic platform, because of its penetration of the Internet. This is clearly a demand-oriented monopoly and this issue will require a policy of providing diverse knowledge in various language environments to maintain cultural diversity. However a problem can be found not just in diversity but also unification.

9.6.3 Competition and Coexistence of Cultural Elements

The frameworks of diverse formality and the interpretation of culture and a set of actions and values are platform goods and can have value if they are shared. What kind of structure does the mechanism of competition and coexistence regarding this platform called culture and the various cultural emblems that make up this platform have? When we were embedded in the cultural platform as a single integrated package and regarded it as a given of everyday life world, the individuals embedded in this life world did not have any choice relating to this package. However, things have changed significantly. There are many structural elements of culture.

If these elements are to be chosen separately, we are going to have to live in an unstable life world where we are exposed to constant choice pressure. Our spiritual world up to modern times would have had difficulty in dealing with such complexity. Culture can be a stable platform for a shared pattern of world views and actions if it is combined, even though it varies. Many everyday life elements thus normally become clustered and the clusters themselves become units of evaluation and choice. Elements themselves inside the cultural clusters are replaced even more slowly.

This clustered platform is on its way to being distinct from maintenance and control of the nation state today. On the other hand, for many everyday life world elements, a de facto standard common to the whole world will be gradually formed and it will be shared and come to function as a platform for new activities, just like business protocols in market transaction, for example. As the formation of world religions gave a platform of common value to a world of tribal units, some kind of platform to be shared needs to be formed in a structure that is not vertically integrated with a specific culture.

In this platform, we will face several different issues such as the maintenance of diversity of individual cultures, the elimination of exclusivity and the deregulation of choice pressure on individuals. To address these issues, there is no choice but to establish diverse cultural, organizational and social networks, which serve as medium units that allow multifactorial belonging, and to construct a place where these networks refer to each other, evaluate, learn from and select each other. Organizations and institutions remove us from the complexity of direct choices as individuals. Organizations and networks are places where the objects of these various everyday life worlds are formed and maintained. The next question is how to construct a higher place where these networks themselves can refer to each other and are evaluated and selected multifactorially. This question cannot be answered just with the market mechanism. To answer this question, a new resource allocation mechanism needs to emerge, for example the accumulation of knowledge by information network technology, reference and circulation mechanism and electronic donations on the network.

9.6.4 Design of a New Civilization

In the world today, there is a flow to dissemble the cluster of culture on the one hand so as to enable a recomposing of the dissembled elements. On the other hand, there is a direction to help reconstruct the everyday life of individuals who are exposed to the changes in these dissembled cultural elements and to design them as a system.

A movement towards vertical integration of cultures on civilization platforms, such as a onetime Christian civilization or Islamic civilization, was accompanied by an exclusive structure, which is on a collision course with other platforms. Conflicts associated with the same kind of exclusive structure are still occurring even today on many cultural platforms such as traditional religion platforms, ethnic platforms and national platforms. This must be something that is already out of time. The crossover of cultural elements is being steadily generated by the Internet. However, a different type of vertically integrated structure is rising on the linguistic platform simultaneously with the evolution of information networking. This is the so-called English Imperialism. It is a new structure of cultural comparative advantage that grew out of the share effects (externality) as a linguistic platform. In information civilization, it is necessary to avoid a lock-in to a specific cultural platform or language from the mechanism of the market and the network.

The 21st century is apparently an era of change with a number of cultural cocoons that used to wrap us already on their way to losing their validity in the midst of our turbulent times. It will take more time for local cultures to regain a new stability in the new civilization called information civilization. Until then, we have no choice but to get to know the nature of the game well, live with this game and design the system. The advanced information civilization is one of the inevitable terminus ad quem of civilized society. This new civilization may appear to be something far different from the culture as our arch-root, but we cannot do anything but to evolve toward this.

The nation-state used to be a system to create artificial people. However, the recognition being shared nowadays is that the nation-state cannot be a platform for controlling diverse cultures. What our new civilization requires is diversification of traditional and unique cultures on the one hand and a condition on the other hand that our everyday life world would not be vertically integrated with a specific cultural platform and the individuals would never be embedded in it. We must discuss how to design relaxed ethical regulations and the minimum platforms common to the network civilization that satisfy the above two conditions and what changes are required for traditional cultures so that the diverse cultural platforms with traditional roots can coexist with and allow easy crossover.

A system where a specific organization monistically controls culture should not last any longer. Even the nation-state is not exceptional in this term. In Europe, the independence of individuals first arose as a form of Protestantism and became the foundation of later economic development. Further, the system of the nation-state itself is losing its meaning in the sense of value control. However, a movement to vertically integrate people's life world is still active even today in the form of religion and ethnicity.

Traditional culture also needs to have a kind of temporalizing process associated with economic growth in order to coexist with multifactorial values. This temporalization was generated in the form of economic growth in Japan and in Western Europe. In Japan, it was supported by the climate of Asian syncretism. In India, for example, the caste system is weakening amid temporal economic development. In Islamic civilization, on the other hand, we cannot say that economic growth and its associated temporalization are evolving rapidly, except in Turkey and Malaysia. Economic growth comes first or cuts across temporalization and its growing influences, take many different forms in world's cultures.

The world that allows coexistence of multifactorial cultures subsumes not just marketing values, but also cultural value formation and the existing economy of donation. This world should not take the form of "non marketability = non competitive environment." Anything that forms social value, except for preserved historical items, is required to have learning and evolutional dynamism by the sharing of knowledge and multifactorial evaluation. For this purpose, we need to construct non-marketable evaluation and choice and logics and a system for resource allocation in an appropriate form, and these will be the new civilization platform. We need to design new social circuits for evaluation and choices based on the multifactorial values and for resource allocation. It is essential the information civilization platform that enables the coexistence and crossover of multifactorial culture becomes a system core common to all our global cultures just as business protocol has become in the market.

In essence, just as the natural sciences have a base of a Universal Law applying to the physical world, in the social sciences a set of Universal Principles applying to the economic and social worlds we have constructed, would be of immense value in ensuring the continuing rich diversity of cultures on our planet.

References

[A]

[Ackoff, 1981] R.L. Ackoff, *Creating the Corporate Future: Plan or Be Planned for*, John-Wiley, 1981

[Andersen, 1996] Esben Sloth Andersen, *Evolutionary Economics—Post-Schumpeterian Contributions—*, Pinter Publishers, 1996

[Aoki, M., 1998] Masanao Aoki, *New Approaches to Macroeconomic Modeling: Evolutionary Stochastic Dynamics, Multiple Eqilibria and Externalities as Field Effects*, reprint edition, Cambridge University Press, 1998

[Aoki, 2001] Masahiko Aoki, *Toward a Comparative Institutional Analysis*, MIT Press, 2001

[Apostel, 1961] L. Apostel, Towards the Formal Study of Models in the Non-Formal Sciences, In: *The Concepts and the Role of the Model in Mathematics and Natural and Social Sciences*, edited by B.H. Kazemier and D. Vuysie, Reidel, 1961

[Arai, 1998] Kiyoshi Arai, Hiroshi Deguchi, *Gaming Simulation* (in Japanese), JUSE Press Ltd., Tokyo, 1998

[Archer, 1995] Margaret S. Archer, *Realist Social Theory: The Morphogenetic Approach*, Cambridge Univ. Press, 1995

[Archer, 1996] Margaret S. Archer, *Culture and Agency*, revised edition, Cambridge Univ. Press, 1996

[Arthur, 1994] W. Brian Arthur, *Increasing Returns and Path Dependence in the Economy*, University of Michigan Press, 1994

[Arthur, 1997] W. Brian Arthur, *The Economy as an Evolving Complex System II*, edited with Steven Durlauf and David Lane, Addison-Wesley, Reading, Mass., Series in the *Sciences of Complexity*, 1997

[Ashby, 1956] Ashby, *An Introduction to Cybernetics*, Chapman & Hall, 1956

[Axelrod, 1997] Robert Axelrod, *The Complexity of Cooperation: Agent-Based Models of Competition and Collaboration*, Princeton University Press, 1997

[Axelrod, 1999] Robert Axelrod, Michael D. Cohen, *Harnessing Complexity: Organizational Implications of a Scientific Frontier*, The Free Press, 1999

[B]

[Barr, 1986] Avron Barr, Edward Feigenbaum, *The Handbook of Artificial Intelligence*, Addison-Wesley, 1986

[Bennet, 1980] P.G. Bennet, "Hypergames: Developing a Method of Conflict", Future, pp.489–507, 1980

[Bonabeau, 1999] Eric Bonabeau, Marco Dorigo, Guy Theraulaz, *Swarm Intelligence: From Natural to Artificial Systems*, Oxford University Press, 1999

[Borgers, 1997] Tilman Borgers, "Learning Through Reinforcement and Replicator Dynamics", J. of Economic Theory, Vol.77, pp.1–14, 1997

[Buckley, 1967] Walter Frederick Buckley, *Sociology and Modern Systems Theory*, Prentice Hall, 1967

[C]

[Carley, 1994] K.M. Carley, M.J. Prietula (eds.), *Computational Organization Theory*, Hillsdale, N.J., Lawlence-Erlbaum Assoc., 1994

[Checkland, 1985] P.B. Checkland, *Systems Thinking, Systems Practice*, John-Wiley, 1981

[Checkland, 1990] P.B. Checkland and J. Scholes, *Soft Systems Methodology in Action,* John-Wiley, 1990

[Christensen, 1997] Clayton M. Christensen, *The Innovator's Dilemma*, Harvard Business School Press, 1997

[Conte, 1997] Rosaria Conte, Rainer Hegselmann, Pietro Terna (eds.), *Simulating Social Phenomena*, Springer-Verlag, 1997

[Cornes, 1996] Richard Cornes, Todd Sandler, *The Theory of Externalities, Public Goods, and Club Goods*, 2nd edition, Cambridge University Press, 1996

[D]

[Deguchi, 1984] Hiroshi Deguchi, "Functional Analysis over Exchange Algebras", Sixth-Seventh Seminars on Applied Functional Analysis: Information Theory and Related Topics, Tokyo Institute of Technology, 1984

[Deguchi, 1986] Hiroshi Deguchi, B. Nakano, "Axiomatic Foundations of Vector Accounting", Systems Research, Vol.3, No.1, pp.31–39, 1986

[Deguchi, 1990] Hiroshi Deguchi, "Exchange Algebras and Economic Field", In: Complex Systems with Decision Makers, edited by Masaya Yamaguti, Kokyuroku, Vol.809, pp.88–105, Research Institute for Mathematical Science, Kyoto University, 1990

[Deguchi, 1992a] Hiroshi Deguchi, "Network Decision Making and Genetic Algorithm", International Conference on Economics, Management and Information Technology 92 Proceedings, pp.235–238, The Japan Society for Management Information, 1992

[Deguchi, 1992b] Hiroshi Deguchi, "System Model for International Comparisons of Subcontract Industries" (in Japanese), *Survey and Research of the Machine Industry 3-3*, pp.47–56, Industrial Research Center, April 1992

[Deguchi, 1994a] Hiroshi Deguchi, *Network* (in Japanese), JUSE Press Ltd., Tokyo, 1994

[Deguchi, 1994b] Hiroshi Deguchi, "Network Oriented Change of Industrial Structure" (in Japanese), Journal of the Japan Society for Management Information, Vol.2, No.3, pp.51–71, 1994

[Deguchi, 1996] Hiroshi Deguchi, "Multi Agent Economics and Its Gaming", In: *Modeling and Control of National and Regional Economies 1995*, A Postprint Volume of the 1995 IFAQ/IFIP/IFORS/SEDC Symposium, edited by Vlacic et al., pp.269–274, Pergamon Press, 1996

[Deguchi, 1998] Hiroshi Deguchi, Agent Based Approach for Social Complex Systems—Management of Constructed Social World, In: *Community Computing and Support Systems*, edited by Toru Ishida, LNCS 1519, Springer, pp.62–77, 1998

[Deguchi, 2000a] Hiroshi Deguchi, "Norm Game and Indirect Regulation of Multi Agents Society, Computational Analysis of Social and Organizational Systems" CASOS Conference 2000, pp.92–95, Carnegie Mellon, 2000

[Deguchi, 2000b] Hiroshi Deguchi, *Fukuzutsukei toshiteno Keizaigaku (Economics as a Complex System)* (in Japanese), JUSE Press Ltd., Tokyo, 2000

[Deguchi, 2001] Hiroshi Deguchi, Takao Terano, Koichi Kurumatani, Toshiyuki Kaneda, Hiroyuki Matsui, Taro Yuzawa, Akio Sashima, Yusuke Koyama, Hao Lee, Masato Kobayashi, "Virtual Economy—Agent Based Modeling & Simulation of National Economy—", In: *Agent-Based Approaches in Economics and Social Complex Systems*, edited by Akira Namatame, Takao Terano, Koichi Kurumatani, pp.198–207, IOS Press, 2001

[Deguchi, 2003a] Hiroshi Deguchi, Masato Kobayashi, Hae Lee, "Agent Based Modeling of National Economy—Centralized Mode Analysis of Virtual Economy by GA Simulation—", IEEE CIRA2003, CD Proceedings, 2003

[Deguchi, 2003b] H. Deguchi, "Learning Dynamics in Platform Externality", Section 9, In: *Applied General Systems Research on Organizations*, edited by K. Kijima, R. Sato, S. Takahashi, Springer-Verlag, 2003

[Deguchi, 2003c] Hiroshi Deguchi, "Agent-Based Modeling Meets Gaming-Simulation: A Perspective to the Future Collaborations", ISAGA 2003 Proceedings, pp.1173–1182, 2003

[Dell, 1998] M. Dell, "Collaborating in a Connected Economy", Oracle Magazine, January/February, pp.24–25, 1998

[Dosi, 1991] G. Dosi, "What Is Evolutionary Economics?", Journal of Evolutionary Economics , Vol.1, No.1, pp.5–8, 1991

[E]

[Eigen, 1971] M. Eigen, "Self-Organization of Matter and the Evolution of Biological Macromolecules", Naturwissenschaften, Vol.58, pp.456–523, 1971

[Epstein, 1996] J. Epstein, R. Axtell, *Growing Artificial Societies*, The MIT Press, 1996

[Esben, 1996] Esben Sloth Andersen, *Evolutionary Economics: Post-Schumpeterian Contributions*, Pinter Publishers, 1996

[Espejo, 1996] R. Espejo, W. Schuhmann, M. Schwaninger, U. Bilello, *Organizational Transformation and Learning*, John Wiley & Sons, 1996

[F]

[Flood, 1991] R.L. Flood, M.C. Jackson (eds.), *Critical Systems Thinking: Directed Readings*, John-Wiley, 1991

[Freuder, 1994] E.C. Freuder, A.K. Mackworth (eds.), *Constraint Based Reasoning*, MIT Press, 1994

[Friedman, 1991] Daniel Friedman, "Evolutionary Games in Economics", Econometrica, Vol.59, No.3, pp.637–666, 1991

[Friedman, 1999] Daniel Friedman, "Evolutionary Economics Goes Mainstream: A Review of the Theory of Learning in Games", http://econ.ucsc.edu/~dan/, March 2, 1999

[Fudenberg, 1998] Drew Fudenberg, David K. Levine, *Theory of Learning in Games*, MIT Press, 1998

[G]

[Gamson, 2000] William A. Gamson, *SIMSOC: Simulated Society, Participant's Manual*, 5th edition, Free Press, 2000

[Gaylord, 1998] R.J. Gaylord, L.J. D'Andria, *Simulating Society: A Mathematica Toolkit for Modeling Socioeconomic Behavior*, NY, Springer-Verlag, 1998

[Gilbert, 1994] Nigel Gilbert, Jim Doran (eds.), *Simulating Societies: The Computer Simulation of Social Phenomena*, UCL Press, 1994

[Gilbert, 1995] Nigel Gilbert, Rosaria Conte (eds.), *Artificial Societies: The Computer Simulation of Social Life*, UCL Press, 1995

[Goldberg, 1989] D.E. Goldberg, *Genetic Algorithms in Search, Optimization, and Machine Learning*, Addison Wesley, 1989

[Greenblat, 1981] Cathy Stein Greenblat, *Principles and Practices of Gaming-Simulation*, Sage Publications, 1981

[Greenblat, 1988] Cathy Stein Greenblat, *Designing Games and Simulations, An Illustrated Handbook*, Sage Publications, 1988

[Greenspan, 1996] A. Greenspan, "Mr. Greenspan Looks at the Impact of Technological Change on American Society", BIS Review, pp.1–3, No.113, October 21, 1996

[H]

[Habermas, 1989] Juergen Habermas, *The Theory of Communicative Action, Volume 2*, Beacon Press, 1989

[Haken, 1978] Herman Haken, *Synergetics: An Introduction*, Springer, 1978

[Hardin, 1968] G.R. Hardin, "The Tragedy of the Commons", Science, Vol.162, pp.1243–1248, 1968

[Hayashi, 1989] K. Hayashi, *Networking Jouhou Shakaino Keizaigaku (Economics of Networking Society)* (in Japanese), NTT Press, 1989

[Hempel, 1970] Carl G. Hempel, *Aspects of Scientific Explanation and other Essays in the Philosophy of Science*, The Free Press, 1970

[Holland, 1975] J.H. Holland, *Adaptation in Natural and Artificial Systems*, The University of Michigan Press, 1975

[Holland, 1995] J.H. Holland, *Hidden Order: How Adaptation Builds Complexity*, Addison-Wesley, 1995

[Hsu, 1995] Hsu, Cheng, *Enterprise Integration and Modeling: The Metadatabase Approach*, Kluwer Academic Pub, 1995

[I]

[Ijiri, 1990] Yuji Ijiri , "Momentum Accounting and Triple Entry Bookkeeping: Exploring the Dynamic Structure of Accounting Measurements" (Studies in Accounting Research No. 31), Amer. Accounting Assn., 1990

[K]

[Kaelbling, 1991] L.P. Kaelbling, *Learning in Embedded Systems*, MIT Press, 1993

[Kagel, 1995] John H. Kagel, Alvin E. Roth (eds.), *The Handbook of Experimental Economics*, Princeton University Press, 1995

[Kampis, 1991] George Kampis, *Self-Modifying Systems in Biology and Cognitive Science: A New Framework for Dynamics, Information and Complexity*, Pergamon Press, 1991

[Kandori, 1993] M. Kandori, G. Mailath, R. Rob, "Learning, Mutation and Long Run Equilibria in Games", Econometrica, Vol.61, pp.29–56, 1993

[Katz, 1986] Michael L. Katz, Karl Shapiro, "Technology Adoption in the Presence of Network Externalities", Journal of Political Economy, Vol.94, No.4, pp.822–841, 1986

[Kelly, 1999] Susanne Kelly, Mary Ann Allison, *The Complexity Advantage: How the Science of Complexity can Help Your Business Achieve Peak Performance*, McGraw-Hill, 1999

[Kijima, 1996] K. Kijima, "Intelligent Poly-agent Learning Model and Its Application", Information and Systems Engineering, Vol.2, pp.47–61, 1996

[Kirman, 1995] Alan Kirman and Mark Salmon (eds.), *Learning and Rationality in Economics*, Blackwell, 1995

[Klabbers, 1996] Jan H.G. Klabbers, "Problem Framing Through Gaming: Learning to Manage Complexity, Uncertainty, and Value Adjustment", Simulation & Gaming, Vol.27, No.1, pp.74–92, 1996

[Koshimura, 1976] Shinzaburo Koshimura, *Matrix Economics* (in Japanese), Dobunkan, Tokyo, 1976

[Kuhn, 1970] Thomas S. Kuhn, *The Structure of Scientific Revolutions*, The University of Chicago, Chicago, 1970

[L]

[Lakatos, 1983] I. Lakatos, *The Methodology of Scientific Research Programmes*, Cambridge University Press, 1983

[Langton, 1989] C.G. Langton (ed.), *Artificial Life*, Addison-Wesley, 1989

[Langton, 1990] C.G. Langton (ed.), *Artificial Life 2*, Addison-Wesley, 1990

[Langton, 1994] C.G. Langton (ed.), *Artificial Life 3*, Addison-Wesley, 1994

[Lee, 2000] Hao Lee, H. Deguchi, "Technological Innovation of High-tech Industry—Agent Based Simulation with Double Loop Learning—", International Symposium on Knowledge and Systems Sciences: Challenges to Complexity (KSS2000) Proceedings, pp.224–229, JAIST, 2000

[M]

[Mattessich, 1995a] Richard Mattessich, *Critique of Accounting: Examination of the Foundations and Normative Structure of an Applied Discipline*, Greenwood Pub Group, 1995

[Mattessich, 1995b] Richard Mattessich, *Foundational Research in Accounting: Professional Memoirs and Beyond*, Chuo University Press, 1995

[Mesarovic, 1970] M.D. Mesarovic, D. Macko, Y. Takahara, *Theory of Hierarchical Multilevel Systems*, Academic Press, 1970

[Mesarovic, 1989] M.D. Mesarovic, Y. Takahara, *Abstract Systems Theory*, Springer-Verlag, 1989

[Michael, 1992] Masuch Michael, Warglien Massimo (eds.), *Artificial Intelligence in Organization and Management Theory: Models of Distributed Activity*, North-Holland, 1992

[Mitchell, 1969] J.C. Mitchell, *Social Networks in Urban Situations*, Manchester University Press, 1969

[Morgan, 1986] G. Morgan, *Images of Organization*, Sage Publications, 1986

[N]

[Nelson, 1982] R. Nelson, S. Winter, *An Evolutionary Theory of Economic Change*, Harvard University Press, 1982

[Nicolic, 1996] G. Nicolic, I. Prigogine, *Self-Organization in Nonequilibrium Systems*, Wiley, 1977

[Nonaka, 1995] I. Nonaka, H. Takeuchi, *The Knowledge Creating Company: How Japanese Companies Create the Dynamics of Innovation*, Oxford University Press, 1995

[O]

[Ossowski, 1999] Sascha Ossowski, *Co-Ordination in Artificial Agent Societies: Social Structures and Its Implications for Autonomous Problem-Solving Agents*, Springer-Verlag, 1999

[P]

[Prietula, 1998] M.J. Prietula, K.M. Carley, L. Gasser (eds.), *Simulating Organizations: Computational Models of Institutions and Groups*, CA, Morgan-Kaufmann, 1998

[Pyatt, 1985] Graham Pyatt (ed.), *Social Accounting Matrices: A Basis for Planning*, World Bank, 1985

[R]

[Reich, 1991] Robert B. Reich, *The Work of Nations*, Alfred A. Knopf, Inc., 1991

[Rosen, 1974A] Robert Rosen, "Some Temporal Aspects of Political Change", Int. J. General Systems, Vol.1, pp.93–103, 1974

[Rosen, 1974B] Robert Rosen, "Planning, Management, Policies and Strategies: Four Fuzzy Concepts", Int. J. General Systems, Vol.1, pp.245–252, 1974

[Rosen, 1978] R. Rosen, *Fundamentals of Measurement and Representation of Natural Systems*, North Holland, 1978

[Rosen, 1985] Rosen, *Anticipatory Systems: Philosophical, Mathematical and Methodological Foundations*, Pergamon Press, 1985

[S]

[Salanie, 1998] Salanie, *The Economics of Contract: A Primer*, MIT Press, 1998

[Salmon, 1984] Wesley C. Salmon, *Scientific Explanation and the Causal Structure of the World*, Princeton University Press, Princeton, New Jersey, 1984

[Samuelson, 1997] Larry Samuelson, *Evolutionary Games and Equilibrium Selection*, MIT Press, 1997

[Sato, 2002] H. Sato, H. Matsui, I. Ono, H. Kita, T. Terano, H. Deguchi, Y. Shiozawa, "U-Mart Project: Learning Economic Principles from the Bottom by both Human and Software Agents", *New Frontiers in Artificial Intelligence*, pp.121–131, Lecture Notes in Computer Science, Springer-Verlay, 2002

[Schulz, 1994] U. Schulz, W. Albers, U. Mueller (eds.), *Social Dilemmas and Cooperation*, Springer-Verlag, 1994

[Selten, 1991] R. Selten, "Evolution, Learning and Economic Behavior", Games and Economic Behavior, Vol.3, pp.3–34, 1991

[Shaun, 1995] P. Shaun, Hargreaves Heap, Yanis Varoufakis, *Game Theory: A Critical Introduction*, Routledge, 1995

[Simon, 1957] H.A. Simon, *Model of Man*, John Wiley & Sons, New York, 1957

[Simon, 1979] Herbert A. Simon, *Models of Thought, New Haven*, Yale University Press, 1979

[Simon, 1982] Herbert A. Simon, *Models of Bounded Rationality*, The MIT Press, Cambridge, 1982

[Simon, 1996] Herbert A. Simon, *The Sciences of the Artificial*, 3rd Edition, MIT Press, 1996

[Smith, 1982] John Maynard Smith, *Evolution and the Theory of Games*, Cambridge University Press, 1982

[Suppe, 1977] F. Suppe, *The Structure of Scientific Theories*, 2nd edition, Univ. of Illinois Press, Urbana, 1977

[T]

[Takadama,1999] Keiki Takadama, Takao Terano, Katsunori Shimohara, Kouichi Hori, Shin-ichi Nakasuka, "Making Organizational Learning Operational: Implication from Learning Classifier System", Journal of Computational and Mathematical Organization Theory, Vol.5, No.3, pp.229–252, 1999

[Takagi, 1995] H. Takagi, K. Kijima, H. Deguchi (eds.), *Multimedia Jidaino Ningento Shakai—Poly Agent Shakai—(Human and Society in Multi Media Age—Poly Agent Society—)* (in Japanese), JUSE Press Ltd., Tokyo, 1995

[Takahara, 2003] Yasuhiko Takahara, Mihajlo D. Mesarovic, *Organization Structure: Cybernetic Systems Foundation*, Kluwer Academic Pub, 2003

[Taylor, 1978] P. Taylor, L. Jonker, "Evolutionary Stable Strategies and Game Dynamics", Mathematical Biosciences, Vol.40, pp.145–156, 1978

[Terano, 1998] T. Terano, S. Kurahashi, U. Minami, "Artificial World for Social Interaction Studies", Proc. 6th Int. Conf. on Artificial Life (ALIFE VI), pp.326–335, 1998

[Thom, 1975] Rene Thom, *Structural Stability and Morphogenesis*, translated from the French edition, W.A. Benjamin, 1975

[Tully, 1993] Shawn Tully, "The Modular Corporation", FORTUNE, pp.58–65, February 8, 1993

[V]

[Varera, 1979] F.J. Varera, *Principles of Biological Autonomy*, North Holland, 1979

[Vega-Redondo, 1996] Fernando Vega-Redondo, *Evolution, Games, and Economics Behaviour*, Oxford University Press, 1996

[W]

[Watanabe, 1992] Watanabe, "The Advance of Internationalization and the New Move to a Domestic Regional Distribution of Work Structure" (in Japanese), Shoko Kinyu, pp.4–22, July 1992

[Watkins, 1992] C.J.C.H. Watkins, P. Dayan, "Technical Note: Q-Learning", Machine Learning, Vol.8, pp.279–292, 1992

[Weibull, 1995] J. Weibull, *Evolutionary Game Theory*, MIT Press, 1995

[Wonham, 1976] W.M. Wonham, "Towards an Abstract Internal Model Principle", IEEE Transactions on Systems, Man, and Cybernetics, Vol.SMC-6, No.11, pp.735–740, 1976

[Y]

[Yam, 1997] Bar-Yam, Yaneer, *Dynamics of Complex Systems: Studies in Nonlinearity*, Addison-Wesley, 1997

[Z]

[Zeeman, 1977] E. Zeeman, *Catastrophe Theory: Selected Papers*, 1972–1977, Addison-Wesley, 1977

[Zeleny, 1981] Zeleny, *Autopoiesis: A Theory of Living Organization*, North Holland, 1981

Index

MIX
Papier aus verantwortungsvollen Quellen
Paper from responsible sources
FSC® C105338

If you have any concerns about our products,
you can contact us on
ProductSafety@springernature.com

In case Publisher is established outside the EU,
the EU authorized representative is:
Springer Nature Customer Service Center GmbH
Europaplatz 3, 69115 Heidelberg, Germany

Printed by Libri Plureos GmbH
in Hamburg, Germany